ORDINARY **GENOMES**

EXPERIMENTAL FUTURES *A series edited by Michael M. J. Fischer and Joseph Dumit*

Technological Lives,
Scientific Arts,
Anthropological Voices

KAREN-SUE TAUSSIG

ORDINARY GENOMES

Science, Citizenship, and Genetic Identities

Duke University Press
Durham and London 2009

Printed in the United States of America on
acid-free paper ∞

Designed by Heather Hensley

Typeset in Warnock Pro by Keystone
Typesetting, Inc.

Library of Congress Cataloging-in-Publication
Data appear on the last printed page of this book.

For

JON AND EMMA

CONTENTS

ACKNOWLEDGMENTS

There are many people whose energy and generosity went into making this book possible. My deepest debt is to the many individuals in the Netherlands who opened their workplaces, homes, and personal stories to an oddly curious American graduate student. Having promised anonymity, I cannot name these individuals here but hope they know they have my enduring gratitude for their willingness to share their time and thoughts with me. In particular I want to thank the clinicians and researchers who allowed me to observe them in action over many months and who patiently responded to my numerous questions. I enjoyed their company and was moved by their commitment to the health and well-being of their patients. I also appreciate the extraordinary generosity of the individuals and families who were navigating the daily challenges of managing a genetic condition but who nevertheless made time to help me understand genetics in the Netherlands. In the Netherlands I also thank Annemarie Mol, Rosie Braidotti, and Gerard de Vries for providing institutional affiliations, conversation, insight, and food, as well as Jessica Messman, Ruud Hendricks, Peter Peters, and Marc Berg for intellectual engagement and encouragement. Diana and Bart Mens offered hospitality that I deeply appreciated. Paul Dirks was a wonderful friend and guide during my time in the Netherlands. I continue to feel lucky to have shared so much fun with such a warm and generous person. I also thank Margreet Zoutewelle for camaraderie and for guidance regarding language and everyday life.

The faculty in the Department of Anthropology at the Johns Hopkins University and, especially, my advisors in and beyond the department, Emily Martin, Gillian Feeley-Harnik, and Katherine Verdery, Erica Schoenberger, Richard Cone, Sharon Kingsland, and David Harvey, were astonishingly generous with their time and ideas. I am particularly grateful to Emily Martin for her groundbreaking work and intellectual curiosity and for offering a model of mentoring and professionalism that will always serve as a guide for me. During the time I was there I experienced the Hopkins department as a remarkably rich intellectual environment. The faculty's model of scholarly engagement produced a community marked by lively engagement with and among a group of graduate students that I continually feel grateful to have encountered. Sarah Hill, Eric Rice, Paul Nadasdy, Sarah Hautzinger, Eric Mueggler, Kamran Ali, Wendy Richardson, Monica Schoch-Spana, Laury Oaks, Fred Klaits, Pam Ballinger, Hanan Sabea, Vineeta Sinha, Elizabeth Ferry, Elizabeth Dunn, Roger Magazine, Chris McIntyre, Iveris Martinez, Espy Baptista, and Carlotta McAllister all shared their ideas and their lively humor. Eric and Paul deserve special thanks for their willingness to forego more scenic runs for the sake of my clumsy ankles and for delivering ice cream at a crucial moment. Johns Hopkins was, simply, a fun place to be and to think. It was only later on my tour of academic institutions, when I saw how much time could be taken up with performance and empire building, that I learned how fortunate I was to have been in an intellectual community that could be experienced as being all about the ideas.

In Cambridge, Massachusetts, I benefited from the interest and encouragement of Byron and Mary Jo Good, Mike Fischer, Leon Eisenberg, Joe Dumit, and Alisdair Donald. For two years I enjoyed a lively intellectual home in the Department of the History of Science at Harvard, for which I thank Stephanie Kenen, Allan Brandt, and Everett Mendelssohn. During these years I met an energetic group of undergraduates, learning from them as I learned about teaching, and gaining valuable experience that helped sharpen my thinking about genetics as a cultural object.

In Minnesota I thank my colleagues in the Anthropology Department for their support of this project.

I am grateful to my first graduate students, Matt Wolf-Meyer and Jianfeng Zhu. These remarkable students continually reminded me of the value of ideas and of why teaching is so rewarding.

Over the course of writing *Ordinary Genomes* I have deeply appreciated the able research assistance of Meggie Crnic, Emma Kippley-Ogman, and Erin Bonner. Their sharp eyes and enthusiasm were always appreciated.

The economics of research being what they are, I am grateful for the funding that made this project possible. My research was supported by a small grant from the Wenner-Gren Foundation for Anthropological Research and a Dissertation Improvement Grant from the National Science Foundation (grant #SBR-9307563). The writing of the dissertation that preceded this book was funded by a fellowship from the Anthropology Department at Johns Hopkins and by a Charlotte Newcombe Dissertation Fellowship. Additional writing time was supported by the University of Minnesota through a course release from the Anthropology Department, a semester's leave from the College of Liberal Arts, and a University of Minnesota McKnight and Faculty Summer Research Fellowship.

Quite a few people read and offered generous comments on various aspects of this project. I am grateful to my writing group, including Karen Ho, David Chang, Tracey Deutch, Kevin Murphy, Keith Mayes, and Malinda Lindquist. Emily Martin, Richard Cone, Gillian Feeley-Harnik, David Harvey, Deborah Heath, Sharon Kingsland, Robyn Kliger, Ann Meneley, Diane Nelson, Erica Schoenberger, Monica Schoch-Spana, Katherine Verdery, Lynn Morgan, Kim Fortun, Joe Dumit, John Hartigan, Vinay Gidwani, Janelle Taylor, Fred Klaits, Laury Oaks, Mike Fischer, Susan Lindee, David Valentine, Stefan Helmreich, and Jonathan Kahn each offered comments on various parts of this project at different stages. Mike Fortun and Priscilla Wald generously read and commented on the entire manuscript not once, but twice, offering crucial insights at key moments. This book would not have been completed without their enthusiasm for the project. I know that their engagement with the project and the suggestions they made significantly improved the book. I also thank Ken Wissoker and Mandy Earley at Duke University Press for the attention and support they have given this project. Of course, any misinterpretations are my responsibility alone.

For engaging conversations that have sharpened my thinking and helped to develop new ideas I am indebted to a fluid community of friends, anthropologists, and science studies scholars. I consider myself extraordinarily fortunate to count among this group Karen Ho, David Valentine, Hoon Song, Stuart McClean, Jean Langford, Vinay Gidwany, Jennifer Gunn, Susan Craddock, Rayna Rapp, Lynn Morgan, Donna Haraway, Joe Dumit, Stefan Helmreich, Heather Paxson, Hannah Landecker, Chris Kelty, Kim Fortun, Mike Fortun, Kaushik Sunder Rajan, Susan Lindee, Troy Duster, Jonathan Marks, Faye Ginsburg, Mette Nordahl Svendsen, Klaus Hoeyer, Nick King, Jennifer Fishman, Jonathan Kahn, Ann Saetnan, Sandra Lee, Janelle Taylor, Emily Martin, Richard Cone, Linda Hogle, Diane Willow, Jo.E. Rizzo, Michael Montoya, and Erin Koch. I am particularly grateful to Stefan Helmreich for his lovely suggestion of *Ordinary Genomes* for the title.

Closer to home, I am grateful to those who contribute to the pleasures of living in Minneapolis, adding to the vibrancy of everyday life and thus to my ability to think: Our neighbors on Emerson Avenue: Ken, Joan, Anna, and Greta Nelson; Welcome Jerde, Dan and Hannah Berg and the spirit of Julia Berg; Bonnie Gruen and Bill Tolman; Karen, Jeff, and Mira; David V.; Diane and Jo.E.; and the Reading Rebels Book Club families.

I also thank my family for their support through a long process.

Isabelle Salgado, Amy Evoy, and Julie McClymonds have generously offered support at a distance.

On the home front Jonathan Kahn has been remarkably patient in seeing this project to fruition. His quip that he can practically recite the book from memory is only barely an exaggeration. In addition to reading, commenting on, editing, and pacing while helping me work out precisely what I want a particular sentence to communicate, he has taken on more than his share of family duties to facilitate getting this project finally out the door. Emma, whose entire life so far has been shared with the production of this book, deserves special appreciation for her hours of patience and for her pleasure in words. It is in deep appreciation of the joys they share that I dedicate this book to them.

Introduction

SCIENCE, SUBJECTIVITY, AND CITIZENSHIP

The novelty of modern biopolitics lies in the fact that

the biological given is as such immediately political, and

the political is as such immediately the biological given.

—Giorgio Agamben, *Homo Sacer*

Every ethnography has its genesis in the interests and experiences that put an ethnographer on a particular path. The path to this project runs through the Czech Republic, where a set of unexpected encounters introduced me to the idea that public discussions of human genetics in the Netherlands were being framed in terms of national identity. While conducting research in Prague during the summer of 1991, I met a Dutch nurse, Veronica, at a bus stop.[1] Over the course of the summer we became acquainted and occasionally met to have a meal, go for a walk, or attend a concert together. At some point early in the summer Veronica asked me about my work. She appeared baffled by my reference to the anthropology of science and medicine and by the fact that I was conducting research in Europe, telling me that she thought anthropologists studied people in places like Indonesia and South America.[2] I made several attempts to explain the immunology project about science and popular culture I had been working on and would be returning to in the fall (Claeson et al. 1996; Martin 1994; Martin et al. 1997), but she seemed not to understand the relationship of this work to her conception of anthropology. Finally, I tried another example, pointing out that the U.S. government had recently committed to devoting three billion dollars to the Human Genome Project, a joint project of the National Institutes of Health and Department of Energy intended to map the entire sequence of the base pairs making up a typical human genome. I explained that such a project would have enormous social, cultural, political, and economic implications in addition to its potential value for science and medicine.[3] The genetics example immediately made sense to Veronica. "Oh," she exclaimed, "you mean the genetic passport." I had never heard of a genetic passport, so I asked her to explain what she was talking about. Veronica described the genetic passport as something the Dutch ministry of health was considering. She believed that this passport would be an identity card with one's genetic makeup encoded on it and that it was intended for both identification and health purposes.

The concept of a genetic passport intrigued me. It seemed such an explicit example of biopolitics—the modern project described by Michel Foucault of optimizing and managing individuals and populations in the name of producing the "normal." Veronica and I discussed the issue of the genetic passport in several subsequent conversations. Over the course of the summer I had an opportunity to meet several of her Dutch friends who visited her in Prague. I asked each of them about the genetic passport. Each recalled having heard about the idea and seemed concerned about its possible development and the kind of surveillance it implied, although none knew whether or not it was under serious consideration. As an anthropologist interested in the production, circulation, and consumption of scientific knowledge, I was curious about the significance of such a concept and how it fit with broader Dutch social life and practices. The idea of a genetic passport intrigued me because of the questions it raised about the relationship between scientific knowledge of genetics and broader understandings of the body as well as of human and national identity. The idea of a genetic passport thus led me to the Netherlands and generated my initial questions.[4] Why would the Netherlands, or any country, produce an identity card with one's genetic makeup encoded on it? Why would a national health department, rather than a state department or ministry of foreign affairs, develop an identity card called a passport? Why would the Dutch government be developing and Dutch people talking about a new kind of passport in 1991 when, with the expected implementation of the European Union's Maastricht Treaty in 1992, borders within Europe were about to be opened and become, at least bureaucratically, irrelevant? What did the concept of a genetic passport say about people's beliefs concerning the state's ability or right to see with regard to the bodies of its citizens? What might the idea illuminate about biopolitics and the production of the normal in the context of newly emerging genetic knowledge? What did it say about the contemporary relationships among biology, scientific knowledge, identity, and citizenship? Furthermore, since the idea of a genetic passport had not actively circulated in the United States, its salience in the Dutch context raised important questions about how the production, interpretation, translation, and consumption of scientific knowledge might differ from one national context to another.

Seeking answers to my questions, I discovered that the genetic passport was something of a myth. No department of the Dutch government was planning to develop such a thing. The concept of a genetic passport appears to have been introduced in the Netherlands in 1989 by Huub Schellekens, a prominent Dutch cell biologist, in a Dutch television documentary. This three-part documentary, *Beter Dan God* (Better Than God) (Kayzer 1987), examined emerging genetic knowledge in the context of the history of genetics and especially the history of eugenics.[5] At the time of my research the documentary was repeatedly described to me as having been extremely popular, one of the most widely watched television programs in Dutch history. The idea of a genetic passport rapidly became part of popular culture, circulating in conversations among Dutch people and in popular media (magazines, newspapers, radio, television) in discussions about human genome research and its applications in medical and scientific practices.

At first confounded by my discovery that the genetic passport was a myth, I soon grew even more intrigued. No aspect of social life makes sense outside of the social system in which it operates. The circulation of the concept of the genetic passport in the Netherlands is no exception. Neither is the practice of human genetics itself, which is otherwise so readily decontextualized as a symbol of the universal and a sign of the unmarked cosmopolitan. Indeed, anthropologists have long been interested in myths and other narratives because the themes embedded in the stories people tell about themselves and the world offer important clues to the organization of their cultural values.

The readiness with which Dutch people took up the concept of a genetic passport signaled the centrality of emerging genetic knowledge to how Dutch people imagine the future in relation to science, subjectivity, and citizenship and, moreover, indicates that the ways people grapple with such issues will likely differ across time and space. Increased knowledge of the human genome bears directly on Dutch understandings of relations between national groups and between states as well as on internal differentiations within a national population. The idea of a genetic passport thus opens the door to larger issues. It raises questions about how contemporary ideas and practices concerning genes are made meaningful in any particular national context and how people appropriate and transform their beliefs and practices as they work to reproduce valued aspects of social life.

As I began investigating genetics in the Netherlands, I learned that the idea of a genetic passport spoke to Dutch anxieties about genetics related to the history of Dutch experiences of the Second World War. Prior to 1994, when the Dutch government began requiring identity cards, people in the Netherlands had been required to carry them only once before: during the Nazi occupation of the country in the Second World War. But people also link the concept of a genetic passport to the war in other ways. When I asked a Dutch social scientist about the concept, he told me he was not very familiar with the idea of a genetic passport but went on to talk about the wartime deportation of Jews from the Netherlands. He said that in terms of percentages, more Jews were deported to concentration camps from the Netherlands than from any of the other countries from which the Nazis conducted such deportations. He explained that many people believed the Nazis were able to deport so many Jews not because the Dutch were more anti-Semitic than other Europeans but because the Netherlands had always been a highly organized society, and this organization facilitated the Nazis' ability to find Jews in the Netherlands. He then speculated that the genetic passport might be another example of Dutch emphasis on organization and social order. I thus came to see that the genetic passport referenced anxiety not just about social disorder associated with war and occupation, but also about the problem of social order and obedience.

Since the end of the Second World War the Netherlands' national borders may never have appeared so penetrable as they do today in the face of contemporary immigration and the ongoing process of European unification, which has involved the opening of internal European borders. In discussing the unification of Europe in interviews, the Dutch people I spoke with repeatedly voiced concerns about the opening of national borders and diminishing control over who and what comes into and goes out of the Netherlands.

The theme of borders is inextricable from the history of the Netherlands because the very shape of the country has been repeatedly reconfigured by draining, dams, flooding, and erosion.[6] The people I spoke with during my fieldwork were worried less about the possibility of being flooded by water than about the prospect of being inundated by people and things that might dilute cultural identity and national integrity. Although they were virtually unanimous in their enthusiasm for the European Union for its economic potential, they repeatedly articulated unease about the reproduction of Dutch social life. They discussed their anxieties about the possibility of increased immigration in a country that is already one of the most densely populated in the world (Shetter 1987:41). They also expressed fears about the movement of drugs and diseases across increasingly permeable national borders. Finally, they spoke about the importance of maintaining a primary Dutch, as opposed to European, identity, expressing this worry in their attitudes toward the opening of the borders as well as in their desire to hold on to such valued cultural markers as the Dutch guilder, then their national currency, and the Dutch language.

The idea of a genetic passport helped people to crystallize and express their concerns about rapid developments in scientific knowledge and new technologies and their implications for social life in relation to anxieties about what it means to be Dutch in light of Dutch history, recent immigration, and European unification. The concept of the genetic passport, like the popular interest in the *Beter Dan God* documentary, signaled that Dutch people had particular stakes in the development of new genetic knowledge. This book is the result of my having followed these signals.

In exploring these issues throughout this book I will argue that contemporary genetic practices in the Netherlands are powerfully shaped by two highly valued Dutch social ideals: first, a desire for ordinariness; and second, a commitment to tolerance. These simultaneously contradictory and interconnected social ideals are always in tension with one another in that they speak both to a deep desire to fit in and to an imperative to accept difference. The contemporary significance of memories of German occupation and Nazi eugenic science during the Second World War further complicates perceptions of genetics in the Netherlands today. Although that history is part of a longer history of articulating Dutch identity in contrast to Germany, in the context of genetics it both raises the specter of eugenics and offers most Dutch people a particularly well-articulated understanding of national difference widely constructed as "tolerant Dutchness" in contrast to "intolerant Germanness." Compounding this history is the emergence of newly disquieting ideas about national identity and the future of the nation, ideas related to European unification that also shape the ways many Dutch people perceive contemporary genetic knowledge and practices.

The Netherlands is a place where ordinariness is highly valued, but, at the same time, any attempt to engineer it is met with apprehension and may be challenged by the social imperative to tolerate difference. The history of the Second World War and contemporary anxieties about European unification expand the grounds upon which genomics may be challenged or embraced. The emergence of genetics as big science at the end of the twentieth century, including the possibility of ensuring the production of ordinariness through genetics, has thus created both possibility and anxiety. Indeed, genomics offers a seemingly scientific basis for determining ordinariness while at the same time precisely opening up the arena where it has historically most obviously been in conflict with tolerance.

In the chapters that follow I trace these issues through the everyday experiences of Dutch people as they encounter genetics in their personal and professional lives. By observing this complex mix we will see how biology, citizenship, and identity are inseparably intertwined. The Dutch case thus illustrates how genetics can be a site for articulating national identity and, in so doing, it demonstrates how such phenomena are incorporated into genomics as it is integrated into daily life.

Finally, another central argument of this book is that the field of genetics —refracted through this complex cultural and historical mix—illuminates how scientific practices are deeply tied to the local even as they signal the position of the Netherlands as a quintessentially Western nation that supports and engages in modern, cutting-edge international scientific endeavors. In this sense, the selection of the Netherlands contributes to a vital reconfiguration of what can constitute the field in anthropology, involving an expansion of anthropology beyond its more traditional field sites.[7] In my book I go a step further to challenge an ongoing tendency to homogenize the West (Carrier 1995; Gupta and Ferguson 1997). The practices that produce this tendency are diverse. In anthropology, where geography has been at the center of the discipline since its inception, this tendency is visible in the frequency with which analyses are framed, either explicitly or implicitly, in comparison with underanalyzed notions of the West. In science studies, research focusing on clinical or laboratory practices often does not even consider the West as a category of analysis, leaving differences within the reified West uninterrogated.[8] Challenging the assumptions embedded in these practices is central to my project.

The inclination to reify the West as a monolithic construct is intensified in matters pertaining to science and its implied universality because the power of science itself comes from its ability to obscure its locality. Genetics represents precisely the kind of social practice that people understand as being fundamental to defining the West. People perceive acceptance of scientific principles as a sign of sharing in a common Enlightenment heritage that marks one as modern, cosmopolitan, and thus a member of Western society.[9] To characterize Dutch genetic practice as locally specific is not to suggest in any way that it is somehow a less rigorous or otherwise inferior type of science or medicine.[10] My point here is to demonstrate the specific local interanimation of science, medicine, and culture in a nation otherwise viewed as a part of a monolith of Western modernity.[11] Concomitantly, we will also come to see that in the age of the European Union and of international genetic medicine, what exactly it means to be Dutch is itself in transformation. By focusing on the lived experiences of Dutch people as they work in and encounter genetics in their daily lives, we can see how ideas about the normal, citizenship, and Dutch identities emerge within a dynamic social field of specific cultural values. Thus, in the local practice of the global enterprise of genetics in the Dutch context, the notion that there is any such thing as a monolith of Western modernity or of genetics dissolves.

CONTEXTUALIZING GENOMICS

Since the late 1980s those of us who live in much of the industrialized world have witnessed a major shift in scientific priorities. This shift can largely be glossed as a move away from cold war preoccupations with defense-related science—especially physics—toward a contemporary focus on the life sciences, particularly genomics and bioinformatics, and their potential applications in medical practices. The shift places the body at the center of scientific inquiry both as the source of research materials (DNA, family histories, etc.) and as a site for intervening in biological processes. Biology emerged as so-called big science in 1988, when the U.S. government unprecedentedly made funds available for life science research by funding the Human Genome Project. Its simultaneous emergence with the ascendance of neoliberal market values combined with a largely private, profit-driven American health care system and global pharmaceutical industry perhaps ensured that biology also would become big business. Today, the still-expanding biotechnology industry is a highly capitalized, multibillion-dollar segment of the world economy.[12]

The contemporary significance of expanding knowledge in the life sciences means that examining the production, circulation, and consumption of scientific knowledge and medical practice is an essential component of understanding the dynamics through which power is produced and exerted in relation to persons and bodies in contemporary social life. At the beginning of the twenty-first century people in a wide range of national and cultural contexts are being offered a powerful new theoretical understanding of life and the relatedness of all species, including the human species, as well as a set of practical interventions into human biology and reproduction that aim to exert new kinds of control over the production of persons. In those places where genetic knowledge and its associated technologies are socially available and culturally salient a wide range of people have participated in the production of highly mediated discussions of utopian and dystopian fantasies about genomic futures. At once people are presented with hopeful visions of a future of regenerative medicine supporting longer and healthier lives, while they also express fears of a brave new world of genetic discrimination, new eugenic practices, genetic passports, and increasingly narrow definitions of what constitutes an acceptable member of the human race.[13] The concept of the genetic passport plays on just this double vision. It was imagined as a card that could facilitate what is now called personalized medicine—the idea that genetic knowledge could facilitate physicians' ability to tailor people's medical treatment to the specifics of their genomes. At the same time, a new identity card with detailed information about a person's biology raised the specter of increased surveillance and social control.

The history of the Netherlands easily establishes the country within a paradigm of the scientific heritage of the Enlightenment. Since at least the times of the Dutch pioneers of science Christian Huygens (1629–95) and Anton van Leeuwenhoek (1632–1723) Dutch society has fostered and embraced scientific pursuits and international scientific collaborations.[14] Today, the Netherlands maintains a highly developed and highly sophisticated cohort of scientific and medical experts who actively participate in an international community of genetic researchers and practitioners. Through its national health care system, the Dutch state also maintains an extensive network of genetics centers where genetic information and services are efficiently delivered to broad sectors of the population.

During my fieldwork it became clear to me that contemporary genetics opens up a space in which people produce diverse narratives about this new knowledge. There is no single story about genetics in the Netherlands; its meanings are contested and negotiated within and across social domains. For most Dutch people, genetics evokes a range of potent and sometimes ominous associations. These stories most immediately involve the dynamics of ordinariness and tolerance, the legacy of the Second World War, and contemporary concerns about boundaries between persons, groups, species, and nations. For a laboratory geneticist, however, genetics also means using scientific and technological processes to read the body from the level of chromosomes and DNA. This story is largely one of using genetics to decode the body. For a clinical geneticist, genetics involves diagnosing, classifying, and counseling patients and their families. This is a story that revolves around solving complicated medical problems and highlights the fluidity among the worlds of science, medicine, and everyday life. For patients, genetics involves negotiating the meaning of the body with scientific and medical professionals as well as with those in their various social networks. This story implicates myriad complex social and personal issues ranging from making reproductive decisions to construing the impact of genetic diagnosis upon identity. Of course, these subject positions are not mutually exclusive. They are themselves complex and crosscutting in that individuals may simultaneously occupy more than one subject position in any given context. An individual might at once be a Dutch clinician or a Dutch person who becomes a patient when pregnant or when a family member seeks some kind of genetic testing. Thus, in any specific context a particular individual may bring multiple perspectives to how they understand the meaning and significance of genetic knowledge and practice.

THE ANTHROPOLOGY OF SCIENCE AND MODERN LIFE

The social transformations set in motion by new genetic knowledge in the Netherlands are deeply connected to social processes associated with modernity. Foremost among these processes in my analysis are the interrelated phenomena of biopolitics, normalization, and citizenship. The genetic passport directly raises the question of why Dutch people used a concept so closely associated with citizenship and national identity to express their concerns about new genetic knowledge. Passports, of course, are a state-sponsored instrument for defining and bounding national identity. Genetics operates similarly at a different scale to define and normalize biological identity at the molecular level. A national passport answers the question of who counts as *legally* Dutch; the idea of the genetic passport speaks to the question of who is *biologically* Dutch. In this one example, then, we see biopolitics writ large.

Biopolitics operates most powerfully through its delineation of the contours of the normal. Much of contemporary human genetics involves conceptualizing the body at the molecular level as a means of understanding and producing representations of human normality. Indeed, the Human Genome Project aimed to produce a representation, in the form of a map, of a single, archetypically normal human genome. The idea of normal is threaded through genetics in both theory and practice and is prevalent in the molecular genetics of the Human Genome Project and genetic research and practices focused on human health.[15]

But the normal is a slippery category. Since the normal is produced only in relation to the abnormal (Canguilhem 1991), it is always twinned with the abnormal or the pathological.[16] This twinning is repeatedly visible in discussions of the Human Genome Project, in which one of the central images of the rhetoric its proponents employed was "the idea of a base-line norm, indicated by 'the human genome'" (Keller 1992:294). In mobilizing this image of a norm, these geneticists thus defined health "in reference to a tacit norm, signified by '*the* human genome,' and in contradistinction to a state of unhealthy (or abnormality)" (Keller 1992:295). There also is a tension in the concept of the normal involving its characterization, on the one hand, as "an existing average" (which can be improved upon) and as a "figure of perfection to which we may progress" (Hacking 1990, cited in Keller 1992:298). The ambiguities embedded in the concept of normal are confounded in the context of contemporary molecular genetics, where at the same time that the concept of the normal plays a central role, in fact variation is the norm.[17]

In the Netherlands the scientific process of genetic normalization exists in constant dialogue with powerful, salient local Dutch social practices of normalization. Dutch people frequently express attitudes about the importance of normality in the concept of ordinary (*gewoon*), a positively valued concept constantly employed in daily life. The common saying, "*Doe maar gewoon, dan doe je al gek genoeg*" (Just act/do/be ordinary, then you are already acting/doing/being strange/crazy enough), for example, expresses the emphasis on ordinariness in Dutch social life. In chapter 1 I explore how contemporary Dutch society is imbued with the concept of *gewoon*: it is used in everything from television advertisements to comments parents make in their attempts to encourage good behavior in their children. These processes of normalization locate genetic practices in the cultural and ideological specificity of the Netherlands, where the ability to *do* ordinary can easily be understood as consonant with *being* Dutch. Thus, in the Dutch context the normal and the ordinary work simultaneously as Foucauldian regulatory ideals and also as explicit terms people use every day to articulate cultural values. By recontextualizing the production and application of new genetic knowledge in this way, we will see how taken-for-granted the structures of normality are in everyday life and in biomedicine.

The chapters that follow explore the force of culture in the processes through which people are working out what counts as normal as they encounter genetics in daily life. They reveal the ways the future is being worked out in the present.[18] As we see Dutch researchers, clinicians, and others working out this future we will see that what counts as normal, even in the domains of science and medicine, is highly unsettled. The Dutch case shows not that people accept the idea of a single norm but that contemporary genomics is a space in the present where people are establishing the tolerable limits of human biological variation for the future. In so doing they are also imagining the tolerable limits of citizenship.

Citizenship, understood as a form of belonging codified in law and experienced through the formation of national identity, lies at the core of the modern state.[19] Questions about biology and citizenship raised by my encounter with the idea of a genetic passport were at the heart of my entry into this project. Many citizenship and nation-building projects have had a biological component, referencing such concepts as blood and race (Balibar 1991; Heath et al. 2004; Rose and Novas 2005). This link between biology and nation suggests that questions of citizenship would inevitably surface in the context of new knowledge about biology.[20]

The complexities and contradictions of twenty-first-century citizenship are broadly inflected with concerns about the biological (see especially Heath et al. 2004). This complex terrain has eugenic potential but is also a site of "new forms of power, knowledge, and embodied discipline, along with novel rights and responsibilities" that Deborah Heath, Rayna Rapp, and I describe as "genetic citizenship" (2004:152).[21] In the chapters that follow I invoke the concept of citizenship both to talk about rights and responsibilities and also to explore those aspects of citizenship that involve often unarticulated cultural frameworks of belonging, including notions of biology, that connect specific individuals to a specific state, nation, and identity. As relations between citizen and state become refracted through the prism of genetics, this new biological knowledge suffuses the genomic futures both of individuals and of the nation.

FIELDWORK AND METHODOLOGY

This book is based on a year of multisited ethnographic fieldwork in the Netherlands, conducted between August 1993 and August 1994. My primary field sites included one of the eight genetics centers in the Netherlands, two high schools, and a community center for people over the age of fifty-five (a category of people known in the Netherlands as *vijf-en-vijftig plussers*). In addition to participating in and observing activities in these sites I conducted more than 110 in-depth interviews.

The field is never as bounded as the delineation of clinic, high school, and community center might imply. I also sought out a number of people who had specialized knowledge about genetics in the Netherlands: a director of a center for bioethics; the staff of an umbrella organization coordinating activities of the support groups representing various genetic disorders; a historian and a cell biologist who coauthored a history of eugenics; the producer of a ten-part television documentary about genetics that aired while I was in the Netherlands; and a staff person from the Society for the Protection of Animals involved in a campaign against genetic manipulation in animals.

I also learned much from participating in the routines of daily life and the numerous informal conversations and experiences such activities entail. Though it is impossible to detail the innumerable daily occurrences that ultimately inform the way one comes to understand the meanings embedded in daily activities, some of the experiences I had stood out for the way they informed my understanding of the Netherlands and of genetics. I remember encountering Dutch bureaucracy in the process of applying for a residence permit; I seemed to be the only American among what seemed to be over a hundred people likely of Turkish and Moroccan origin who, like myself, were seeking the right to be in the country. I traveled widely by both bicycle and public transportation, taking in the landscape, the weather, and the efficiency with which one is able to get around the country. Shopping at local stores and outdoor public markets, going to museums, attending an open house at a local community center; engaging in social activities with a wide variety of people, sharing an apartment with a Dutch university student, and participating in holiday activities such as the arrival in the Netherlands (by boat, from Spain) of Sinterklaas (St. Nicholas or Santa Claus) with his Moorish assistant Zwaarte Piet (Black Piet) all informed my understanding of the Netherlands. Such open-ended experiences of everyday life are a central component of ethnography. They also provide an indispensable framework for broadly contextualizing specific practices of genetics in the Netherlands.

Over the course of any given week I would find myself involved in a diverse set of activities through which I came to understand how genetics was embedded in everyday life in the Netherlands. Most mornings I would ride my bike to the genetics center, where I would join one or more clinicians as they examined and consulted with patients or sit in on meetings of clinicians as they worked to diagnose the puzzling array of symptoms they had encountered in particular patients during the previous week. Some days I would interview a clinician or laboratory geneticist. Several days I observed the local obstetricians performing amniocenteses, listening to the interactions between and among the medical specialists, the pregnant woman, and the person(s) accompanying her (usually her partner) as the physician used a needle to draw amniotic fluid from the woman's belly in order to grow and examine the fetal cells contained in the fluid. I would regularly join one of the clinicians at the local teaching hospital, where they spent the afternoon meeting with pregnant women, typically with their partners, who had been referred for consultation with a geneticist. In these meetings I would listen while the geneticist used pictures of cells and chromosomes to explain cell division, conception, and genetic risk. Over a series of Monday evenings I made sure I was home in time to watch *Een Rondje* DNA (A Round of DNA)—a ten-part Dutch television program that aimed to educate people about genetic knowledge. One evening while watching the program I was struck by a description of cell division, conception, and genetic risk that was similar to that given by the geneticist at the hospital. Rather than using the pictures of cells and chromosomes I had seen in the clinics, however, the program used cooked spaghetti noodles to represent genetic material for their intended popular audience. I attended the genetics segment of a biology course at a high school and volunteered at a community center for those over fifty-five. I asked everyone I encountered—clinicians, researchers, patients, students, retirees, people I met in the course of everyday living—if I could interview them about their experiences with genetics. Many evenings I would take the train or ride my bike to the family home of an individual or, more often, a couple who had generously agreed to share their thoughts about their experiences with genetics. Some of those I spoke with had little formal experience with genetics, having primarily considered it in relation to things they had read or seen in the media. Some were pregnant and being confronted with the possibility of genetic testing, information about risk related to genetic conditions, and the potential to make choices about whether to carry a pregnancy to term. Others were caring for a family member, usually a child, with a genetic condition or were themselves living with such a condition. Still others were engaged with genetics in their work lives as clinicians or researchers.

During my fieldwork in the Netherlands my access to information was uniquely shaped by my distinctive position within the structures of Dutch society. For example, geneticists, above all, those working in clinical practice, repeatedly expressed surprise that an American, especially one from Johns Hopkins University—a major research university with a highly developed medical genetics program—would come to the Netherlands to study genetics. My institutional affiliation granted me a status in their eyes that facilitated my access to information about their activities.

All ethnographic stories are necessarily partial because of the temporal constraints of fieldwork. I was in the Netherlands for only one year, and the people I encountered had stories that began before I arrived and continued after I left. The partial nature of fieldwork was made clear in the context of the genetic centers where I conducted much of my research. The services the centers provide often involve multiple encounters that take place over many months and even years. Even the cases I followed from the first to what was considered the final clinical appointment often remained open-ended with the recommendation that a child who had been the subject of the encounter return for genetic counseling when she or he began thinking about having children.

OVERVIEW

This book takes on the challenge of recontextualizing science by moving from the broad contours of Dutch society and history to the particularities of contemporary genetic practices in specific domains. Three primary themes repeatedly emerged during my fieldwork on genetics. They are the subject of my first chapter. The Netherlands is characterized by a deeply segmented social structure that promotes and reinforces an ideal of a self-consciously tolerant society. The Dutch, I learned, enable a social ideal of tolerance by bounding and containing difference so as to minimize its social threat. It is within this ideal of tolerance, I argue, that Dutch people construct and manage the meaning of genetic difference. Yet Dutch values about ordinariness and fitting in limit Dutch tolerance. It is impossible to live in the Netherlands, moreover, and miss the powerful and persistent significance of the legacy of the Second World War. Attitudes toward genetics in the Netherlands are deeply informed by Dutch associations of genetic practices with the Nazi program of racial hygiene. This legacy has influenced how Dutch people I met both inside and outside of the genetics centers make sense of genetics today. Chapter 1 thus lays the foundation for a fuller understanding of the broad social dynamics surrounding genetic knowledge and practice in the Netherlands.

In Chapter 2 I examine contemporary discourses about the significance of new genetic knowledge and the institutional framework through which Dutch people encounter that new knowledge. It is within and in interaction with these highly integrated institutional networks, involving clinics, research and diagnostic laboratories, and departments dealing with psychosocial problems associated with genetics, that human genetic knowledge and practices are produced and reproduced in the Netherlands. Nowhere is the production of that knowledge more evident than in the centers that form the basis of my inquiry in this chapter.

Specific Dutch encounters with genetic knowledge and practice are the focus of my investigation in the subsequent chapters. In chapter 3 we see the most vivid illustrations of normalization at work in the dynamic production of genetic diagnoses within and beyond the clinic. I argue here that the clinical practice of genetics in the Netherlands is produced in a convergence of medical practitioners' desire to identify and pathologize difference and the tensions embedded in Dutch commitments to both tolerance and ordinariness.

In chapter 4 I turn to religion, showing how the Dutch medical geneticists with whom I worked engage ideas and values about the relationship among religion, biology, and the geographical location of religious communities in developing clinical diagnoses. I also explore how religion in the Netherlands functions in the popular imagination as a means of understanding the dynamic through which local ideas about geography, social practice, biology, and religion converge with scientific knowledge in producing clinical interpretations of the body.

My final chapter is a case study of a nonprofit group that gained national attention as it articulated concerns about developments in genetics. This chapter highlights the centrality of the body and identity in popular constructions of the significance of genetics. It also demonstrates that these constructions develop out of contestations over the meaning of the body and identity within and across various social domains, as is clear in the startling visual images of human-animal hybrids that highlight the production and dissemination of specific interpretations of the meaning and value of genetic practices.

Throughout this book readers will come to see the complex processes through which local culture and scientific and medical practices mutually engage, contest, and inform each other. Together, these processes will highlight the multiple ways the local production of scientific and medical knowledge of genetics and its application in practice intertwine with history, religion, geography, and political economy to produce Dutch identities. In so doing, these chapters ask one to rethink prevailing concepts of modernity, science, and the West. For as one comes to see how embedded genetic knowledge and practices are in everyday life in the Netherlands, one realizes that genetics cannot be extracted from its national and historical contexts. Research that does not take into account these contexts either misses an important part of the story or relies on uninterrogated assumptions about modernity, science, and the West.

Chapter One

"GOD MADE THE WORLD AND THE DUTCH MADE HOLLAND"

In Europe we are not so concerned with one's work. . . .
You know for us the character of a person matters more.

—quoted in Scott Haas, *Are We There Yet?*

An elaboration of some aspects of Dutch history and social life will contextualize my analysis of practices and beliefs surrounding genetics in the Netherlands. Three major themes frame this analysis. The first involves how Dutch society in general deals with difference through social structures that accommodate it and social values that tolerate it. I approach the question of difference historically by examining the Dutch tradition of bounding religious difference in institutional structures that render it less threatening. How Dutch society deals with difference in general directly shapes how Dutch people both inside and outside the clinic construe and manage genetic differences in particular. The second theme involves considering how a constraining aspect of Dutch tolerance produces a powerful dynamic of normalization. Together with their attitudes of tolerance, Dutch people simultaneously emphasize the importance of fitting in by being ordinary or typical. The third theme speaks to how the legacy of the Second World War affects contemporary attitudes toward both difference and genetics. Many Dutch people strongly feel the need to define their society against Nazi intolerance toward difference as manifested in the Nazi program of racial hygiene. This elaboration of Dutch social life analyzes quotidian knowledge to illustrate how cultural values often thought to be so obvious as not to warrant examination in fact demand the closest anthropological attention in order to show their generative force.

The title of this chapter is a Dutch saying. It refers to the fact that much of the territory of the Netherlands is made up of land reclaimed, through human effort, from the ocean. The Dutch literally made the provinces of Holland as well as parts of the other provinces making up the Kingdom of the Netherlands, much of which is below sea level. I use the title here to stress that the Dutch, like any society, also make their social world.

TOLERANCE AND THE MANAGEMENT OF DIFFERENCE

A central aspect of contemporary Dutch identity involves the idea that Dutch people are self-consciously tolerant.[1] Dutch tolerance grows out of a history of religious diversity and is supported by social structures that help manage difference at a societal level. I begin with a discussion of the Oscar-winning Dutch film *Antonia's Line.* The film uses Antonia's story as a celebration of a widespread ideal of modern Dutch personhood as tolerant, secular, and antifascist. I then elaborate on aspects of Dutch social life that contextualize the ways Dutch people interpret and deal with difference. These include the Dutch history of religious pluralism, its institutionalization in the structure of what Dutch scholars describe as *verzuiling* and translate as "pillarization" (referring to the idea that Dutch society stands on several distinct "pillars" or social groups), and the more recent process of secularization. This analysis reveals that contemporary Dutch identities are deeply informed by the ideal of tolerance.

Antonia's Line

In 1996 the Dutch movie *Antonia's Line* won the American Oscar award for best foreign film. Although the film received mixed reviews in both the Netherlands and the United States, it is significant as a celebration of important aspects of contemporary Dutch social life. The film valorizes the daily enactment of a Dutch social ideal of tolerance. It also emphasizes the importance of ordinary middle-class life, the persistent significance of the Second World War, and the secular nature of life in the Netherlands today.

Like any text, a film can be read in numerous ways. *Antonia's Line* centers on the lives of four generations of one family. It is, perhaps primarily, a feminist film about strong, outspoken, self-sufficient women. This feminist story is one of autonomy from men. Against the backdrop of this theme the film also presents an account of broad changes in social life since the end of the Second World War that resonates resoundingly with contemporary understandings of the recent past and with key aspects of contemporary Dutch identity.

The film opens and closes on the day of Antonia's death. In between it reviews Antonia's life from the end of the war, when she returns to her ailing mother's farm in her natal village with her own daughter, Danielle. In opening with explicit reference to the end of the war, the film indexes a key moment in Dutch history. It marks the beginning not only of Antonia's story but also of a new era in modern Dutch life. One might say that Antonia's story is an idealized story of the contemporary Netherlands.

The mother soon dies, and Antonia and Danielle, apparently the only surviving family members, settle into running the family farm and into a variety of relationships with people in the village. As a woman and single mother farming her land on her own, Antonia is a bit of an oddity in the Netherlands of the 1940s. Her status as both a landowner and native to the village means she is deserving of acceptance from other villagers, no matter how grudging that tolerance might be. It is clear from the beginning of the film that other villagers view Antonia as not appropriately conforming to the norms of village life. Yet, she is also undeniably a native who belongs there.

The primary members of Antonia's social circle are all decidedly out of the ordinary. They include

— the village idiot, who loyally follows Antonia after she chastises a group of young children who were taunting and throwing rocks at him;
— Deedee, the developmentally disabled daughter of the local wealthy family, whom Danielle brings to live with Antonia after finding her being raped by one of her two brothers;
— Farmer Bas, a widower, and his three sons, who do not have good relationships with most village residents because, having lived in the village for only twenty years, they are considered outsiders;
— Olga, a Russian émigré who runs the village cafe/pub;
— "Crooked Finger" or "Finger," an eccentric intellectual who is an old friend of Antonia's. Along with some of Antonia's other village friends, Finger was active in the Resistance during the Nazi occupation. He now never leaves his home.

Each of these characters represents a specific type of difference illustrative of popular ideas about the problematic nature of village life before the war and of the negative side of Dutch social life. The village idiot is a social problem; Deedee's mental retardation is a genetic problem. Farmer Bas and Olga, as outsiders, represent a geographical problem, in particular the problem of immigration. Finger is a problem because his principled behavior during the war stands as a reminder of other villagers' acceptance and possible complicity with fascist authority. Each of the characters is a problem because each stands out as individually distinctive from the norm.

Nevertheless, the film portrays the characters in Antonia's circle in a positive light as colorful and of value. The film depicts the bulk of the village population, in contrast, as boorish, colorless, and backward. The film draws its negative portrayal of the mainstream villagers by highlighting their distance from Antonia and their intolerance of perceived outsiders. It also focuses on their apparently blind and hypocritical religious devotion and raises questions surrounding their potential complicity with the Nazis. Taken together, these images serve to represent the villagers as outside the contemporary Dutch ethos of secular, tolerant, antifascism as represented by Antonia. Antonia's circle is also distinguished from the local wealthy family, which is marked by the domineering presence of men and the greed and excess associated with wealth and unchecked power.

The film celebrates Antonia's generous tolerance by showing how it enriches her life and the lives of those around her. It contrasts Antonia's colorful life full of family and friends with the drab lives of the mainstream villagers and the bitter lives of the local elite. The viewer sees Antonia leading a happy and fulfilling life surrounded by people whom she knows and cares about and who know and care about her. Tolerance in the form enacted by Antonia, which involves acceptance, caring, and understanding, reflects widespread attitudes about what tolerance in its ideal form involves in the Netherlands today. Antonia's moral strength is further demonstrated in her outspoken criticism of the local priest and other villagers for their intolerance of outsiders, the hypocrisy of their religious practices, and their complicity with the Nazis during the war.

As the years pass, the circle around Antonia and Danielle expands. Nevertheless, those making up the circle continue to be marked by their difference from the traditional rural Dutch population. In contrast to the earlier members of Antonia's circle, who represented the problem of conventional Dutch (especially village) life, the new characters represent differences associated with modernity. These differences include sexual orientation, increased opportunity and autonomy for women, single parenthood, and secularization.[2]

The film highlights the rapid secularization process in the Netherlands since the 1960s through a series of changes in village life. As time passes, fewer and fewer villagers are active in the church, while those who are die. One priest leaves the church because he finds it too repressive and therefore stifling of who he really is as a (Dutch) person. He becomes part of Antonia's social circle and eventually marries the single mother of two who has also joined Antonia's circle. Meanwhile, a Protestant man and Catholic woman who had been prevented from marrying across religious lines die futilely of broken hearts.

The film represents Dutch attitudes about the problems of wealth by contrasting Antonia's richly fulfilling middle-class life and values to the selfishness and cruelty exhibited by the family of wealth. Deedee finds a haven at Antonia's farm, where she happily marries the sweet village idiot. Meanwhile, her family destroys itself through its greed and excess. One of Deedee's brothers leaves the village after raping her and joins the military. He returns years later upon his father's death only to collect his inheritance. While in the village he again commits a rape, this time of Antonia's preadolescent granddaughter. In perhaps the ultimate test of her tolerance, Antonia, when offered an opportunity for revenge, spares his life. He is left to meet his fate at the hands of his equally cruel brother, who kills him to secure a larger share of their inheritance.

The film also appears to recognize the persistence of some of the most extreme aspects of the conservatism of past social life. One could, for example, read Farmer Bas as a symbol of contemporary foreigners in the Netherlands—newcomers who are not well integrated or well accepted. The movie delivers the message that a good Dutch person such as Antonia is tolerant of newcomers. In other words, Antonia represents the ideal modern, enlightened Dutch person who lives a largely secular middle-class life, speaks up about injustice and hypocrisy, and is tolerant and accepting of those who are different from her or who are outsiders.

Ultimately, *Antonia's Line* celebrates the value of living a secular, middle-class, modern life informed by morally appropriate acceptance, understanding, and tolerance of difference. This version of tolerance is one that recognizes difference without rendering it hierarchical. During my fieldwork in the Netherlands I found these values manifest in an emphasis on recognizing shared humanity in spite of difference. This attitude is most clearly reflected in the way Dutch people appear to think of all people as the same but different.

The Same But Different

The theme of tolerance and conceptualizations of all people as the same but different persistently emerged in a wide range of settings during my fieldwork. Understanding how Dutch people manage difference is essential to understanding the ways genetic differences are handled socially, scientifically, and medically in the Dutch context. Dutch values about tolerance and associated understandings of similarity and difference are defining features of Dutch identity, constantly being produced and reproduced in daily life. They shape the ways Dutch people I met both inside and outside of clinical settings conceptualize Dutch identity and interpret and react to difference, including differences associated with genetic disease and abnormality. Recognition of the significance of pluralism and understandings of tolerance in the Dutch context are so pervasive that those studying the Netherlands almost inevitably point to them as crucial features of Dutch social life.[3]

Scholars who study the Netherlands recognize the long existence of religious diversity and tolerance dating back to the Reformation. The long history of religious tolerance and heterogeneity in the Netherlands resonates with contemporary values, central to Dutch identity, which produce and maintain social norms while tolerating difference. Dutch tolerance since the Reformation is striking in comparison with the religious oppression of other European countries over the same period. Indeed, religious diversity in the Netherlands was to some extent the direct result of the immigration of religious groups such as the Flemings, Huguenots, and Jews who were fleeing persecution in their countries of origin.[4] A number of authors have discussed the presence of religious tolerance and the social forces that supported it in detail for the Dutch Golden Age (Gouden Eeuw), the period in the seventeenth century during which the Netherlands became a global power, trading in spices and diamonds. The Dutch sociologist Johan Goudsblom remarks that Dutch merchants' "commercial interests were better served by peace and tolerance than by theological zealotry." He goes on to argue that these concerns moderated the influence of Calvinism and "created a social climate favorable to freedom of thought, for which the Dutch Republic was famous throughout the seventeenth and eighteenth centuries." As a result, he argues, "Dutch society, besides offering sanctuary to many foreign refugees, could preserve a varied religious composition among its native population" (Goudsblom 1967:18; see also, e.g., Schama 1988:61–62, and van Deursen 1991:233).[5]

Religious Heterogeneity in the Netherlands Today

Tolerance of social difference in the Netherlands is rooted in the country's history of tolerance, understanding, and acceptance of religious pluralism —attitudes that have been institutionalized in the Netherlands since the mid-eighteenth century in the structure of verzuiling.

The Dutch sociologist J. P. Kruijt points out that "it is no accident that the only language which has a name for . . . a [social] bloc is the Dutch language. In our country we call such a bloc a zuil. . . . The word pillar is a rather apt metaphor." He explains that

> a pillar . . . is a thing apart, resting on its own base (in our case a particular religious or non-religious faith) separated from other pillars . . . they are standing upright, perpendicular sets of persons and groups separated from other sets. Perpendicular means that each pillar is cutting vertically the horizontal socio-economic strata which we call social classes. For a pillar is not a social class; it contains persons out of every social class or stratification. We might say that the "horizontal functional integration" is crossed by the "vertical ideological integration." Further, a pillar is solid; the ideological pillars of the Dutch nation are indeed strong super-organizations. . . . Finally, all the pillars together generally serve as a support to something resting on top; in our case, that something is the whole Dutch nation. At least, that is what is signified by this word pillar. (Kruijt 1974:129–30)

In the Netherlands, religious groups built a pillarized infrastructure to draw sharp boundaries around different devotional communities. These boundaries were also drawn around other identifiable social groups, such as socialists and humanists, as they emerged. Historically, these boundaries have strongly influenced individuals' experiences of Dutch social life.

This influence was evident in a lecture presented by the Dutch historian D. J. van der Veen that introduced the concept of pillarization to a group of foreign college and university students enrolled in a summer program in the Netherlands. Van der Veen described his experience of growing up within the system of pillarization in the 1950s:

> I am Protestant by birth and when I talk to my Catholic peers it seems to us that we are from different countries. . . . When I was born my mother was helped by a Protestant midwife and my birth was announced in the Protestant newspaper. The announcements (and paper) were printed by a Protestant printer. . . . I went to a Protestant school . . . we didn't go to the greengrocer next door, who was Catholic, because we imagined that the quality was no good and the prices exorbitant, but rather, we went several blocks away to the Protestant greengrocer where they had exactly the same things but we believed that the quality and the prices were far better . . . we went to Protestant summer camps . . . and followed the Protestant t.v., radio, and newspapers. (van der Veen 1994)

Van der Veen went on to explain that when he discussed his education with his Catholic peers they found that the lessons and textbooks used at their denominational schools provided quite different interpretations of historical events. It did not seem surprising, for example, that Catholic schools would teach a very different analysis of the Netherlands' relationship with Catholic Spain than would Protestant schools. Van der Veen also explained the social structure of pillarization as "living apart together." Echoing Kruijt's analysis, he argued that the pillars forming the structure were "holding up a common roof" and said that the idea of living apart together shaped the organization of the Netherlands into the 1960s.

Kruijt's analysis of the pillarization of social life thus reflects van der Veen's experiences. Both stress just how deeply religious affiliations reached into both the organization of Dutch social life and the production of a wide range of social institutions. In discussing Roman Catholics and orthodox Protestants, Kruijt contextualizes pillarization in the Netherlands by listing how the practice engenders parallel institutions serving the respective groups:

> The two groups have not only their own schools, their own political parties, their own press, but also in general:
>
> - Their own trade unions, farmers' unions, employers' unions, shopkeepers' unions, co-operatives, agricultural loan banks;
> - their own institutes for social research and societies for physicians, for lawyers, for teachers, for social workers, for scientists, for employees, for artists, for musicians, for authors;
> - their own music bands, choral societies, sport clubs, theatre clubs,

travelers' clubs, dance clubs, clubs for adult education, "public" libraries, broadcasting;

- their own youth organizations, women's clubs, student clubs, fraternities, and sororities;
- their own hospitals, sanatoriums, organizations for all kinds of social work and charitable work, etc.

I stop without even trying to be complete. (Kruijt 1974)

The pillars of Dutch society crosscut categories of class, hierarchy, region, ethnicity. As Kruijt argues, the pillars are "based on a Weltanschauung (view of life). . . . They exist within a large democratic society (nation), a society that is mixed in terms of Weltanschauung, but that racially and ethnically is predominantly homogeneous" (cited in Post 1989:12).

Pillarization developed in the Netherlands over four centuries, but its contemporary instantiation is understood as stemming from the "schools issue" of 1857 (Shetter 1987:178–86), when both the most conservative branch of Dutch Protestants—the *Gereformeerden*, or Orthodox Reformed—and Catholics demanded state recognition and funding for church schools. These populations, both of which identified as minority groups, were reacting to moves by the dominant Protestant elite to institute education on an increasingly secular basis. The struggle over schools both produced and reinforced religious identity and promoted allegiance to distinct religious values. National political parties emerged to support and advocate for various devotional communities, thus further segmenting Dutch social life (Goudsblom 1967:31). As other groups that shared cohesive values emerged—socialists at the end of the nineteenth century and later a secular group—they fit into the established pattern by forming their own bloc or pillar.

After the Second World War some, especially socialists and liberals, made concerted efforts to do away with the structures of pillarization as well as the practices that reinforced its institutionalization. In reaction to their experiences of war, however, many people articulated a strong desire to reproduce or reinvent what they remembered of prewar social life—to go back to how things had been before the horrors of the war. These attitudes produced a powerful reinstantiation of pillarization, making the immediate postwar era one in which pillarization had an even greater role than it had in its prewar incarnation (Lorwin 1974; Post 1989; Shetter 1987). Dutch schools remained a central location for the production and maintenance of a pillarized society. Since 1920 all schools, including those with religious or private affiliations, have been subsidized and subject to oversight by the national government.[6]

Moving beyond one's community, or bloc, was not without consequences. Jeanine, a seventy-nine-year-old woman from a Protestant background whom I interviewed, described the problems she faced when she married a Catholic man. She said that at the time they married—around the Second World War—she and her fiancé could not find a minister or priest willing to perform the ceremony, so she married in a civil ceremony in her parents' home. She said it was quite a scandal because in the absence of a religious ceremony most people considered them not to be married at all. Although she did take the step of marrying across religious lines, Jeanine's story is reminiscent of the couple in *Antonia's Line* who died of broken hearts because of their inability to take such a step.

In 1967, at what might be considered the end of the hegemony of religious pillarization, Goudsblom suggested that "it may well be that the process of verzuiling has already reached its pinnacle and is now on the decline" (Goudsblom 1967:127). By 1967 a number of cities, among them Amsterdam, and some of the rural areas in the northern Netherlands, much (in some cases more than half) of the population was registered as "unchurchly" (Goudsblom 1967:57). Since 1967 the continuing secularization of Dutch society has been rapid and dramatic. In its wake Dutch sociologists have articulated the concept of *ontzuiling*, or depillarization to describe the declining significance of the institutional structures of pillarized society (Goudsblom 1967; Post 1989).

By the time I arrived to carry out my fieldwork in the Netherlands in 1993 numerous studies had already argued that the Netherlands has become a highly secular society (Middendorp 1991). During my fieldwork Dutch newspapers reported a study suggesting that the Netherlands has the lowest church attendance in Europe, while other supporting studies claim that the Netherlands has the highest number of people who say they do not believe in God. Numerous Dutch churches have been turned over to nonreligious use. During my fieldwork I frequently passed churches that had been converted into apartments.

One of the schools where I interviewed teenagers is formally Catholic, but the significance of this affiliation is played down. As one teacher told me, "This is a Catholic school, but you don't really notice it now." The classrooms were populated by students from Catholic, Protestant, and

Muslim backgrounds as well as by students who said they did not belong to any religious faith or were nonbelievers (*ongelovigen*). The majority of Dutch people no longer marry in church. In contrast to the circumstances Jeanine faced when her marriage took place outside the church, today a marriage in the Netherlands is not legal without a civil ceremony, while a religious ceremony is legally optional.

One evening, while having dinner with Janneke and Rudie, a young married professional couple, I mentioned my interest in the process of verzuiling. They both gave me rather confused looks. Thinking I had pronounced the term incorrectly, I tried again. This time Rudie understood what I was talking about, but Janneke was still confused—even when he repeated the term. After Rudie explained the structure that I was referring to it was clear that this organization of social life was not something that Janneke, who had attended Catholic schools and sings in the choir of the Catholic church she regularly attends, was conscious of, either as a historical phenomenon or as an aspect of her daily life.

In spite of the rapid secularization and the decline in the significance of pillarization, many of the values and beliefs associated with the various religious groups as well as the structures associated with the phenomenon of pillarization persist. For example, the majority of television and radio stations in the Netherlands were associated with one or another pillar. Dutch people are highly attuned to the distinctions among the stations. My roommate, a university student, would frequently mention a station's orientation in discussing a television program, explaining, for example, that such and such a program is broadcast "on the Christian station." The oversight of schools continues to be strongly pillarized even though, as in the case of the Catholic school where I conducted interviews, the students attending a particular school do not necessarily identify with the social bloc in which the school itself is rooted.

Many additional markers silently communicate religious differences. While Protestant tradition leads people from Protestant backgrounds to wear their wedding rings on the third finger of their right hand, Catholics wear their wedding rings on the third finger of their left hand. Names too can be associated with a specific religious background. Jeanine, the woman who could not find a priest or minister to perform her interfaith marriage, told me that in 1993, during a stay in the hospital, the doctor was surprised to see that she had put down on her form that she was Protestant. She said that he asked her about it specifically because "he saw that I had a Catholic last name [her husband's] but that I had put on the form that I was Protestant." In the Netherlands the newspaper delivered to one's home, the stores one shops in, the schools one or one's children attend, the churches one attends, in addition to one's last name and where one wears his or her wedding ring, have historically marked religious differences, one of the primary social differences among Dutch people.

The significance of certain structural aspects of pillarization remains evident in the persistence of some of the basic building blocs that historically formed pillars. If a group can show that it has a point of view for which there is interest in the population, they may be able to gain access to the institutional structures that produce the social blocs making up the pillars. These structures include, among other things, broadcasting licenses and school support. When I was doing my fieldwork there were state-funded Islamic schools and Turkish and Arabic language radio and television broadcasting organizations serving the country's substantial Turkish and Moroccan immigrant populations.

Pillarization as a Worldview

The institutional nature of pillarization operates, at least for many young people, under the radar. It is in this sense that Kruijt's articulation of a relationship between pillarization and a worldview in the context of an otherwise homogeneous society is helpful. The current status of the institutional structure of pillarization is less significant than the fact that its deep embeddedness in Dutch society over many decades forged a particular mode of understanding the world. Pillarization no longer functions as a mighty religious structure but as a worldview that provides a conceptual model for incorporating new or divergent social practices into Dutch social life. Pillarization thus serves as a strategy to construe and manage many forms of difference. The strategy of pillarization (as opposed to its formal structure) allows for the general tolerance of social heterogeneity by bounding difference and minimizing its social threat while containing it within the larger commonality of Dutch society. Beyond religious pluralism, tolerance in the Netherlands at the time of my fieldwork was also oriented around many other forms of difference, including those associated with race, ethnicity, nationality, and sexual orientation. At the same time, however, new immigration, particularly from rural Turkey and Morocco, had, as I elaborate below, already begun to strain the ability of many

Dutch people to incorporate the differences presented by these new social groups into the rubric of the same but different (see, e.g., Abraham-van der Mark 1989).

Emblematic of contemporary Dutch tolerance are attitudes toward differences in intellectual ability and educational achievement. Even as the Dutch provide extensive support for public education from kindergarten through university, they do not necessarily consider academic achievement an essential prerequisite to a full, successful life. As *Antonia's Line* suggests in its portrait of the two developmentally disabled characters, those who do not excel academically are not necessarily stigmatized as being somehow inferior persons.

Education also involves power, and a major function of any educational system is the reproduction of social life, including existing social hierarchies (Bourdieu 1967; Bourdieu and Passeron 1977). The Dutch system is designed to function on merit, but widespread expectations that children will follow the same educational paths their parents did continue to help reproduce prevailing social hierarchies. Nonetheless, while differences in educational achievement translate into money and power, they do not reproduce significant disparities in the status of an individual as a full and valued member of society.[7] And although people certainly recognize class distinctions, they work hard to minimize their effects and widely support the right of everyone, regardless of intellectual or other ability, to access the means to a middle-class lifestyle.

Tolerance of differences in educational ability was made clear to me in a discussion I had with Janneke and Rudie about high school education. Like many European school systems, the Dutch system tracks students from an early age. Janneke asked me about the system of high school education in the United States. I explained that students generally were expected to finish high school, which went through twelve grades, and that students were usually about eighteen at the time they finished. Janneke and Rudie, who are, respectively, a physician and an engineer, both seemed quite surprised that all students would be expected to finish so many grades. I explained that students do drop out and also that there is a test (General Equivalency Diploma, or GED) that students can take to gain the equivalent of a high school diploma. I added that it was widely considered extremely difficult to get a job without a high school diploma, which generally required completing twelve grades. Janneke's response to my comments was striking: "But what if a person is not very good at school, or if they do not like it? Maybe they are good at other things, and they would rather put their energy in that direction. Not everything requires so much education."

I thought about this conversation numerous times during and after my fieldwork because Janneke's surprise seemed so distinct from widespread American discussions about staying in school and about the importance of, especially, math and science education.[8] What Janneke acknowledged in her discussion with me is that people have different talents and that these abilities should be recognized, accepted, understood, and valued. This kind of attitude is facilitated by a worldview rooted in pillarization, which allows Dutch people to recognize some, even some quite stark social differences without necessarily stigmatizing them. In other words, the Dutch tradition of pillarization provides a conceptual strategy for managing difference through tolerance.

Dutch conceptions of tolerance, however, are not synonymous with a "do your own thing" attitude. Blakely (1993:10) argues that difference in the Netherlands has historically been tolerated because differences were constructed as if they were inconspicuous. Pillarization plays a powerful role in rendering certain differences unremarkable by constructing them as part of a known group or category, thereby making them more manageable.

When difference is bounded within a known category such as religion, race, or educational ability, it is widely perceived as deserving acceptance. It is not, therefore, that Dutch tolerance works simply because one is not supposed to notice difference, though this is certainly a component of how it does work. Rather, tolerance works because the bounding of social difference into categories minimizes its social threat. It is, in part, its bounding that makes difference unremarkable. Dutch people today most closely achieve the social ideal of tolerance when difference is bounded and contained within recognized categories that can be understood as Dutch. The increasing racial, ethnic, and religious diversity in the Netherlands, the result of ongoing immigration, particularly from Turkey and Morocco, today is straining Dutch orientations to tolerance in ways I (and many of the Dutch people I know) could never have imagined in the mid-1990s. At that time Dutch people thought that the right-wing political tensions arising elsewhere in Europe—especially in France and Austria—would be unable to gain a foothold in the Netherlands because they were so at odds with the widespread commitment to the Dutch ethos of tolerance and accommodation. The significance of these strains in the Netherlands more recently is perhaps most disturbingly visible in the context of two murders —one in 2002 and the other in 2004—that highlight the contemporary challenges to the Dutch social ideal of tolerance.

Both murders involved increasing tensions around Muslim immigration, above all that from Turkey and Morocco. The Netherlands is frequently identified as having the highest population density in the developed world. *Nederland is vol op* (the Netherlands is full) is a phrase used to express the idea that the country cannot absorb more people. Indeed, one might view the genetic passport's suggestion that there are people in the Netherlands who do not belong as foreshadowing more recent and more hostile sentiment in this regard. During my fieldwork Dutch citizens regularly voiced concerns about the presence of immigrant communities, particularly those made up of Turkish and Moroccan immigrants and their families who entered the country as "guest workers" in the 1970s.[9] Such issues were often expressed in relation to the idea that these immigrants were not learning culturally salient practices, such as speaking Dutch or riding a bike. The failure to adopt these virtually essential aspects of Dutchness made it difficult to incorporate these groups into a rubric that depends upon a concept of the same but different. There simply seemed to be too many differences to make the "same" part of the formulation work. Prior to 2000, there was, nevertheless, a national conversation about the Netherlands now being a multicultural society (something that teenagers especially articulated in conversations with me) and an explicit commitment to recent immigrants that included, for example, the kinds of institutional supports offered to a social group in the context of pillarization—in this case, teaching immigrant children in their own languages in primary school and state support for newspapers and radio and television programming that reflected and expressed the dominant views of the largest immigrant populations.

In May 2002, the anti-immigrant political party founded by Pim Fortuyn came in second in national elections, taking twenty-six seats and forcing a coalition government with the parties that had come in first (the Christian Democrats) and third (the Liberal Party). The widespread shock at the success of the politics of exclusion in the Netherlands was only exacerbated by the shock of Fortuyn's murder five days before the election by an animal rights activist who claimed to view Fortuyn's politics and its successes as a threat to democracy. In fact, there has long been a small right-wing minority in the Netherlands. Nevertheless, historically, any modest electoral successes it had were widely viewed as an affront to the Dutch ideal of tolerance, such that in subsequent elections the extreme right would lose whatever ground it had previously gained.

What makes the Fortuyn story so powerful is not just the extent of his party's success but also the fact that the extremist, violent response *also* challenged Dutch values about recognizing and tolerating difference. And, although Fortuyn's party did not remain in power long, some of its policies regarding immigrants and immigration have been adopted by more mainstream parties, thus highlighting the fact that ideas once perceived as thoroughly at odds with Dutch social ideals have become increasingly accepted.

In January 2004 the Dutch government issued a report stating that the nation's experiment with multiculturalism had failed and calling on immigrants to assimilate immediately (Roxburgh 2004). The report criticizes Dutch policies regarding immigrant populations, arguing that these had created ghettos of immigrants, especially in regard to those from Turkey and Morocco. The report calls on the state to promote policies that will enable the immigrants to "become Dutch" (Roxburgh 2004). The report specifically criticizes these immigrants for tending to marry within their own communities and for failing to learn Dutch, suggesting that these practices, facilitated by prior state policies, "perpetuated their alienation and prevented them from integrating into Dutch society properly" (Roxburgh 2004). In March 2004 the Dutch parliament passed an order to eject twenty-six thousand refugees seeking political asylum. Many of these individuals have been in the Netherlands for more than a decade, and they have children who speak Dutch and attend Dutch schools (Brendel 2004). The debates sparked by these events were framed in relation to the Dutch ideal of tolerance (Brendel 2004).

In November 2004, the Dutch media were again filled with news of extreme and violent death. This time the murder was of the Dutch filmmaker Theo van Gogh, who was reported to have received death threats after having made a controversial film about women in Islam. The work had been shown on Dutch national television. The film was written by Ayaan Hirsi Ali, a Somali-born member of the Dutch parliament who is an outspoken critic of the status of women under Islam and a highly controversial figure in the Netherlands (she too received death threats). Van Gogh was shot and stabbed by a man whom Dutch prosecutors describe as an Islamic extremist; the assailant slit van Gogh's throat and left a death threat to Ali pinned to his body with a knife. Although only one man is accused of the murder, Dutch investigators believe he may well have had help; they arrested and were investigating more than a dozen Muslim men shortly after the murder. Perro de Jong, a Dutch radio commentator, reported to the BBC on Dutch reactions to the murder. In an article titled "Dutch Fear Loss of Tolerance" de Jong writes, " 'Today is the day I became a racist' was one of the typical reactions that appeared on Dutch websites" after the attack, even before the perpetrator was confirmed to be of Moroccan descent (de Jong 2004). Mosques in several Dutch cities were the targets of vandalism and unsuccessful arson attacks in the wake of van Gogh's murder. While the murder has intensified public discussions about the problem of immigrant populations in the Netherlands, many Dutch Muslims were reported among those protesting the killing in the center of Amsterdam, some "carrying banners with slogans such as 'not in the name of my Islam' " (de Jong 2004). One of the major criticisms of the Dutch state in the aftermath of van Gogh's murder was that politicians had ignored the rise of Islamic fundamentalism (de Jong 2004), an extreme orientation vastly at odds with Dutch values of moderation, at least in part because of the ideal of tolerance. Like not speaking Dutch and not learning to ride a bicycle, religious fundamentalism has become another symbol of the contemporary challenge to the Dutch social ideal of tolerance. Part of what is troubling about this challenge to Dutch tolerance is concern about whether the challenges can be dealt with in a way that maintains the integrity of what so many Dutch people value in the ideal of Dutch tolerance.

The Dutch structure of pillarization historically served to manage difference. In its contemporary form as a worldview that produces and reproduces the Dutch social ideal of tolerance, pillarization works to bound and contain difference. The dynamics of the worldview of pillarization shape the way Dutch society deals with difference in general. Ideally, such tolerance involves recognizing, understanding, and accepting difference without rendering it hierarchical. I explore the way tolerance operates vis-à-vis genetic differences in greater detail in chapter 3. The bounding and containing of difference, however, have a constraining element in another Dutch social ideal—ordinariness—an ideal that demands conformity within groups.

JUST BE ORDINARY

Above I examined how people in the Netherlands draw on history, social structure, and social values to manage difference. In its contemporary manifestation as a worldview, pillarization can work to bound and contain difference so as to minimize it as a challenge to social tolerance. Pillarization, however, also produces the constraining Dutch social ideal of being *gewoon*, or ordinary. Ordinariness, which emphasizes moderation, typically demands conformity within groups. Pillarization, however, not only demands conformity, but also enables the articulation of group difference. The emphasis on ordinariness suggests that Dutch tolerance does not easily extend to extreme difference or radical individualism. People recognize, accept, and tolerate differences but are most comfortable with those differences when they conform to a socially recognized group or category such as religion or sexual orientation. The pressure to be ordinary emphasizes conformity within groups and facilitates the repression of individual articulations of identity. Through the worldview of pillarization, Dutch people simultaneously work to minimize difference and stress similarity.

The social significance of ordinariness produces a potent dynamic of normalization. Understanding the dynamic of normalization in its broader social context helps contextualize the social specificity of the unusually intense focus on diagnosing and categorizing people in Dutch genetics centers, a process I examine in chapter 3. I argue that pillarization as a worldview, in which both tolerance and ordinariness are simultaneously at work, cannot be dismissed as window dressing on a universal normalizing ideology. Rather, normalization, a central component of modern power, as Foucault understood it, operates differently in distinctive social, historical, political, and economic contexts.

Foucault's understanding of power is useful in studying contemporary social life because of his conception of power as productive. For Foucault, modern power is a function of knowledge developed through discourses of knowledge that produce social norms. He conceives of normalization as a process by which modern subjects internalize norms produced through authoritative forms of knowledge. Although one can understand power, as such, as productive, modern power is also repressive as people internalize and strive to elaborate common social ideals. In the Netherlands, normalizing processes that impose a social ideal of ordinariness cannot be separated from the social ideal of tolerance. This ideal enables the production and reproduction of multiple categories of normal. In the Netherlands tolerance and ordinariness work in tandem as people strive to accommodate diversity and maintain social order. Thus, in the Netherlands normalizing processes not only produce the normalizing ideal of ordinariness, but also contribute to the construction and maintenance of a self-consciously tolerant society. The distinctiveness of Dutch processes of normalization helps illustrate the importance of understanding social processes associated with modernity as locally specific. Conceiving of modernity in this way helps illuminate the problems caused by collapsing separate societies into a single, undifferentiated category of the West.

"Just Normal, in the Middle, Like Us"

As I developed a network of relationships in the Netherlands, people invited me to join them for social activities. Eventually I realized that virtually every time I had a first social encounter with a Dutch person he or she would observe that I did not fit their stereotype of Americans, often asking me to explain why. Sometimes their impressions of Americans were based on firsthand experiences. More often they were founded on perceptions developed from encounters with other Americans in the Netherlands and with Americans in the media, especially television—at the time, the Oprah Winfrey show was enormously popular in the Netherlands.

Take, for example, the conversation I had with Marie, a geneticist in her late twenties. While riding home after spending a Sunday afternoon together, she mentioned that "we [Dutch people] imagine that everything in America is big." I laughed and said that in America we often think that everything in Texas is big. Marie continued in an amused voice: "We imagine that the cities are big, that the cars are big, that the houses are big, the drinks are big . . . that the people are big, but you're not big, why aren't you big?" After I stumbled over some explanation about being from a family of "small people," Marie went on to tell me that she had heard on the radio that a summer camp in the United States was designed for overweight children: "In the Netherlands we really cannot imagine such a thing!" she said. Both the idea of so many fat children and the existence of a camp organized to help them lose weight were quite odd from her perspective.

When I visited the home of Lieve and Eric, a couple in their late thirties, they told me about a trip they had taken to the United States the previous summer. They thought it was funny that Americans have such huge containers for drinks that are filled mostly with ice and strange that people in grocery stores with their carts filled with bags and bags of food nevertheless stopped to buy and eat greasy fast food. Lieve said, "I thought it was very funny that everybody was either very big [*dik*] or very thin [*dun*]." I did not recognize the word she used for thin, so I asked her what she meant. She switched to English, saying, "Thin, thin, like you" and then, switching back to Dutch, said, "Everybody was either fat or thin but nobody was just normal, in the middle, like us." They went on to say that they think Americans are fat because they eat so much fast food. They explained that they themselves occasionally go to McDonald's but not often and that in general they eat a lot of fresh fruits and vegetables.

Another friend told me that the "typical Dutch diet" includes a lot of fresh fruits and vegetables. Typical in this case is distinguished from traditional, the traditional Dutch diet being much higher in fat and including far fewer fruits and vegetables. People frequently made this distinction in interviews when discussing how their diet today is typical or ordinary but differs from that of their parents and grandparents.

These encounters highlight two types of distinctions many Dutch people make in articulating Dutch identity. The first involves opposing Dutch and other. Oppositions between Dutch and other gain salience as they serve to reinforce notions of Dutch people as tolerant and moderate. Dutch articulations of tolerance, for example, implicitly stand in opposition to widespread understandings of intolerance in Germany and France. People frequently articulate this notion of intolerance with the idea of chauvinism. I regularly asked Dutch people how they thought people in neighboring countries differ from Dutch people. They most often responded to this question by telling me that both the Germans and the French are "a bit more chauvinistic" (*een beetje meer chauvinistisch*). The term *chauvinism* here connotes a polite criticism of what is perceived as overbearing nationalism that construes difference hierarchically.

Dutch perceptions of German intolerance, as I elaborate below and in chapter 5, are represented historically in Nazism and more recently in the contemporary resurgence of neo-Nazism in Germany. People frequently used the example of language to discuss French chauvinism, suggesting that the monolingualism prevalent in France and the intense concern with maintaining the French language reflected an excessively insular attitude of bias against non-French people.

For many Dutch people, German and French chauvinisms are problematic because they construct difference hierarchically. Yet, as the Fortuyn and van Gogh murders and the events surrounding them illustrate, Dutch tolerance also has its limits. Many Dutch people look down upon behavior they perceive to be excessive or extreme. It is precisely the Dutch perception of the excessive nature of German and French chauvinism—and of Islamic fundamentalism—that troubles them. Similarly, as my discussion above illustrates, America is perceived as being materially excessive; although no Dutch person I met would put it quite so baldly, many perceive the United States as a land of many excessively fat people. These attitudes promote and reinforce long-standing Dutch attitudes about the value of middle-class life and the importance of fitting in.

The second type of distinction people make in articulating Dutch identity separates the present from the past. The significance of this type of opposition stems in part from how it articulates a contemporary ethos of modernity in opposition to popular understandings of the recent past. Understandings of the recent past, illustrated in stories that Dutch people tell about themselves, such as *Antonia's Line*, frequently highlight the insular conservatism of social life in the recent past. Although in its contemporary form pillarization as a worldview enables tolerance as a means of managing difference, in what follows I elaborate further on the limits of the Dutch ideal of tolerance. These limits are produced in the contemporary form of pillarization as a worldview through the simultaneous valuing of tolerance and ordinariness.

Constraining Difference—"Just Be Ordinary"

Even as it allows for differences among groups, pillarization as a worldview emphasizes the importance of conformity within categories. This constraining aspect of pillarization is best expressed in the Dutch value of ordinariness. *Doe maar gewoon, dan doe je al gek genoeg* (just do/act/be ordinary, then you are already doing/acting/being strange/crazy enough) is one of the most common sayings in colloquial use. The value of ordinariness imposes conformity in two ways. First, it works to situate all individuals within groups—to bound difference within known categories. Second, within each category it demands conformity to characteristics that are socially understood as ordinary for that group. Understanding the social value of ordinariness in the Netherlands helps illuminate the constraining aspects of the social structure of pillarization.

The concept of ordinary (gewoon) is one that people in the Netherlands use constantly in daily life. Significantly, in the Dutch context the idea of ordinary does not carry the negative connotation that it does at least in American English. Rather, *gewoon*, which I translate as "ordinary" but could also translate as "customary," "typical," or "usual," is a highly valued social ideal. *Doe maar gewoon*—acting and, in so acting, being ordinary is a central organizing principle of Dutch society. The word *gewoon*, which has several common uses, was employed in virtually every interview I conducted. The concept permeates Dutch social life, playing a pivotal role in the intersections among the social production of similarity, difference, and tolerance.

When used in the phrase *gewone mensen* (ordinary people), the term has a class connotation, referring to the working class. People never used the term in this way in interviews with me, and indeed there are more polite ways to talk about class. The other two common uses of the term were employed regularly in interviews. The first of these are cases in which one could translate *gewoon* as "just" or "simply." When a student was explaining to me that people's attitudes about sex and about talking about sex were *gewoon vrij open*, what she was communicating was that they are "just very accepting [tolerant]." This kind of usage is a bit tricky, for it seemed to me that while *gewoon* often translated as "just," I found that "just" in many of these contexts could also mean "ordinary."

Almost invariably people I interviewed employed the term *gewoon* to respond to my questions about whether or not they did anything special for their health, such as eating, exercising, taking vitamins, or anything comparable. Frequently the use of *gewoon* in such contexts meant both "just" and "ordinary":

> **KST:** Is there anything special that you do for your health? . . . [I pause, people almost invariably shake their heads in the negative and then I continue] . . . Maybe with eating, exercise, vitamins, or something comparable?
>
> **Beatrix:** No, we just eat ordinary [*gewoon*], ordinary vegetables, ordinary fruit.
>
> **Frits:** No, we just eat ordinary [*gewoon*], just [*gewoon*] bicycling.

What people communicate in these examples is that they eat fruits and vegetables that would not be considered out of the ordinary; and they are not talking about bicycling for training, racing, or exercise in the sense in which it is used in the United States, but about ordinary, everyday cycling to and from work or school and to run errands.

Tineke, a twenty-seven-year-old university student, explained this kind of usage by telling me that people use *gewoon* in this way to "conclude [a thought] and sometimes also to imply that you think it is weird that people ask you for the reason for something. Like if they would say: 'Can't you understand? It's simple, I just think it is like this.'" Tineke remembers the explosion in the word's use in the matter-of-fact way she describes after *gewoon* was used this way in a peanut butter commercial in the mid-1980s. In that commercial a little boy described the peanut butter as being *gewoon lekker*. In this example we again can see the multiple meanings generated in the idea of gewoon. *Gewoon lekker* could be translated as "just yummy," but, because both *gewoon* and *lekker* are so deeply embedded in colloquial Dutch, the saying implies that the peanut butter is delicious in a way that Dutch people like. Peanut butter is not a traditional Dutch food: one informant suggested that it originally came to the Netherlands via Indonesia, a former Dutch colony. By using *gewoon* the advertisement not only tells Dutch consumers that the peanut butter is tasty, but also communicates the notion that peanut butter is something that typical or ordinary Dutch people enjoy eating. In other words, the peanut butter is delicious specifically in the context of Dutch sensibilities.[10]

Acting gewoon, as one woman in her sixties explained, is associated with moderation, including, for example, eating and consuming in moderation as well as moderating one's behavior so as not to call attention to oneself. She told me that in her family they do everything in moderation (*met mate*). The social emphasis on ordinariness and moderation in the Dutch context, then, also stresses conformity. Several people, in talking about the importance of not standing out or drawing attention to oneself, told me a Dutch saying that emphasizes moderation and conformity: "In the Netherlands all the grass needs to be cut to the same height."[11]

The publication in 1994 of a book of tips for foreigners in the Netherlands suggests the ongoing significance of behavior associated with ordinariness in the Netherlands today. The book, by Hans Kaldenbach, is entitled *Just Be Ordinary: 99 Tips for Interacting with Dutch People* (*Doe Maar Gewoon: 99 Tips Voor het Omgaan met Nederlanders*) and includes suggestions about appropriate social behavior, advice about suitable topics of conversation, and explanations of typical Dutch sayings and behavioral norms. The introduction to the book states that it is "especially directed at people of Turkish, Moroccan, Surinamese, Antillian, and Aruban origins" (Kaldenbach 1994:5); the cover illustration of a man with a dark complexion wearing a fez and patting a lion, a symbol of the Dutch nation, suggests who the book's intended audience is. The tips help illustrate the range of behaviors Dutch people might consider imposing ordinariness upon. One tip explains that Dutch people find it unfortunate if foreigners speak with each other in their own language because it makes the Dutch mistrustful—they imagine you are talking about them (13). Another tip explains that if a Dutch man asks you to give his greetings to your wife, there is not a hidden sexual meaning in the request. Rather, he is simply acknowledging in an ordinary, socially acceptable manner that you have a wife and that she was not present during the social encounter. The book suggests that if a Dutch person offers such greetings to your wife you should respond by thanking him and saying that you will pass on the greetings (22).

Kaldenbach also emphasizes that the Dutch value moderation. He offers a tip on Dutch values when it comes to enjoyment, explaining that "many Dutch people have a problem with enjoyment. It is almost as if they think enjoyment is a sin. It seems that there is an internal inhibition that holds them back from all-out enjoyment or letting themselves go. This is apparent in their moderate eating, their simple clothing, the sober architecture, the subdued manner of celebrating, and so forth" (Kaldenbach 1994:12). Such ideas about moderation are embedded in Dutch commitments to ordinary middle-class life.

A conversation I had during an interview with two teenagers also highlights how the idea of ordinariness orients social life in the Netherlands. The schoolmates Annemiek (who was sixteen years old) and Marc (eighteen) had a lot to say in response to my question about whether or not Dutch people have common qualities or characteristics. When I asked the question they both laughed and said that they think there are a lot of similarities in terms of the way people think (*manier van denken*). Marc said that Dutch people have very accepting attitudes about different cultures, that the Netherlands is a multicultural society. He went on to talk about how Dutch people are tolerant of homosexuality (*denken heel maakelijk over*), alcohol use, and marijuana use. Both talked about how this openness makes the Netherlands unique. Annemiek mentioned tolerant attitudes toward abortion, and Marc followed by mentioning Dutch positions on euthanasia. Annemiek then mentioned that in the Netherlands people are raised to be open, tolerant people (*vrij opvoeding*). When I asked them what they thought about these attitudes, Annemiek said that she thinks it is outstanding (*hartstikke goed*). Marc told me that he thinks it is a very good situation that one can talk openly about things, that one has nothing to fear (*verbergen*). "I think it is fantastic, that that [tolerance] is ordinary [gewoon] in the country. Without that you get shut out, if I say 'I'm homo[sexual].' . . ." At that point Annemiek interrupted with the comment that these attitudes are normal or ordinary, just as they should be (*gewoon vrij*).

Values regarding tolerance and acceptance pervade contemporary Dutch attitudes toward genetics. Such values are directly built into some genetics course materials. A number of Dutch schools employ a lesson plan for teachers of basic genetics that was developed by the Vereniging Samenwerkende Ouder- en Patiëntenorganisaties (VSOP), an umbrella organization of support groups for people affected by genetic conditions. The lesson plan projects broadly held Dutch ideas about similarity and difference onto genetics by titling its first lesson "Different and Yet the Same" (VSOP 1987). By framing genetics in this manner, the VSOP draws upon the Dutch worldview of pillarization to assert the ordinariness of genetic difference.

People in the Netherlands operationalize the concept of gewoon as part of a process of taking something unique or special and making it ordinary. This process can provide a space for Dutch people to see those with genetic anomalies as different and yet the same. Both Annemiek and Marc introduced the idea that the Netherlands is different from other places because of Dutch people's tolerant attitudes about a range of social practices. By the end of the conversation their comments showed not only that they value such attitudes positively, but that such beliefs and practices are appropriate because they are ordinary and, by extension, Dutch.

ARTICULATING SOCIAL DIFFERENCES

The same dynamic by which pillarization bounds difference also minimizes the number of contexts in which individual Dutch people explicitly articulate distinctive social identities. Pillarization enables the repression of such individual articulations because the social structures themselves function as the primary vehicles for articulating identity. In articulating and bounding difference, the structure of pillarization diminishes the need for individuals to do so. Dutch people tend to be comfortable with group designations because so long as a person can be identified with a socially recognized and accepted group or category, she or he may be considered ordinary. If a person were to independently articulate an individual identity, this might threaten the central social dynamic of pillarization through which tolerance and social harmony are maintained. People tend to consider aspects of individual identity a private matter.

I learned about the reluctance to articulate individual difference in two ways. First, I found that people experienced great difficulty in interviews with a set of questions about individual distinctiveness. Second, in numerous interactions with me Dutch people expressed surprise at the kinds of topics Americans ask about and discuss in public settings. Such public discussion of what Dutch people perceived as deeply personal topics served as another symbol of American excessiveness in contrast to Dutch moderation.

Reluctance to talk about topics people perceive as personal has significant consequences for attitudes toward discussing genetic anomalies. I argue below that the reluctance to discursively articulate individual identity helps make individual difference less conspicuous. Inconspicuousness is key to fitting in. The phenomenon of minimizing difference thereby helps people manage it. This is not to say that Dutch people are not self-conscious or do not have fully elaborated senses of individual identity. Rather, their identities are not simply the product of some common modern Western ethos but are distinctively produced through Dutch social life.

In his analysis of the "concept of self and society in the Netherlands," Peter Stephenson criticizes the persistent tendency in anthropological literature to produce a homogenized version of "ourselves"—variously coded as Western, Euro-American, or Industrial—in contrast to "some rather exotic cultural 'other'" (Stephenson 1989:227). Stephenson uses Clifford Geertz's frequently cited essay on concepts of the person as an example of how such homogenization occurs in ethnographic writing: "The Western conception of the person as a bounded, unique, more or less integrated motivational and cognitive universe, a dynamic center of awareness, emotion, judgment, and action organized into a distinctive whole and set contrastively both against other such wholes and against its social and natural background, is, however incorrigible it may seem to us, a rather peculiar idea within the context of the world's cultures" (Geertz 1983:59). This monolithic conception of a modern Western subject is stark in anthropological work that has an explicit comparative intent. Such comments also abound in anthropological literature even when the subject of study is located in the West.

Stephenson asserts that "it remains to be seen just how varied the concept of person in the 'West' might be" (Stephenson 1989:228). He argues that part of what makes people Dutch is a shared, locally specific sense of self. Homogenized references to a Western notion of persons veil such distinctiveness:

> The concept of the self with respect to others in the Netherlands is simultaneously intensely egalitarian and highly individualistic. The compression of these two contradictory polarities—I am just like everyone else/I am unique—meets at the point of persons and draws them intensely toward one another in a cooperative manner while it also fragments them. . . . [I]t is how these Dutch selves, which are rather like ourselves, relate to one another that dramatically differs [from interpersonal relations in other societies glossed as Western]; and this in turn implies a level of subtle difference in the universe within as well. (Stephenson 1989:232–33)

In focusing on the tension between individualism and egalitarianism, Stephenson identifies a key aspect of the way Dutch society manages difference.

Difference in the Dutch context is further mediated, and made inconspicuous, by the Dutch state through its management of the distribution of wealth and access to public resources. Simon Schama's history of the Netherlands emphasizes the deeply bourgeois nature of Dutch society even as early as the seventeenth century (Schama 1988). In their work on legal culture in the Netherlands, the Dutch legal scholars Erhard Blankenburg and Freek Bruinsma conclude by emphasizing the importance of social similarity: "The Netherlands is a flat country, in geographical as well as in social terms" (Blankenburg and Bruinsma 1994:75). The flatness is produced through a highly developed welfare state. The Dutch state maintains a steeply graduated tax system that inhibits the accumulation of great wealth. It also finances and oversees a complex network of social welfare programs, including support for housing, public education, and guaranteed health care. The Dutch sociologist Hans van de Braak points out that the distribution and oversight of these social programs represent about 60 percent of the country's national income (van de Braak 1993:14). Through these social programs the Dutch state actively effaces conspicuous class differences and produces a robust middle-class society. Van de Braak considers the organization of the country's welfare system a reflection of the distinctively Dutch concern of providing for all people. He argues that these concerns are based on the values of freedom and respect for human rights (van de Braak 1993:14).

In my fieldwork I found that Dutch people continually used the idea of being the same, yet different in their daily lives. The simultaneous emphasis on similarity and difference serves to limit expressions of extreme individualism. In the context of genetics, the focus on similarity as a means of minimizing difference affects the way many Dutch people deal with genetic conditions. Their approach to genetic syndromes oscillates between producing a resounding silence around them and an intense urge to categorize them. People repeatedly told me that speaking about genetic conditions has historically been taboo in the Netherlands. A number of people discussed their understandings of the causes of genetic disorders by telling me that historically such disorders have been interpreted as punishments (*bestraffingen*) from God. The shame induced in the Netherlands by this interpretation was understood by at least some informants as having extreme results. One couple told me, for example, that one hears stories of people finding out, after the fact, that a person with a genetic condition had lived their entire life completely hidden from public view. Although the moral and religious values promoting the silence around genetic disorders have diminished with secularization, the silence around such anomalies persists. Several of the physicians in the genetics centers discussed the fact that their patients were reluctant to ask their family members about inherited conditions. The problem of patients not wanting to discuss such problems with their family members also came up repeatedly in clinic meetings.

Dutch people seem much more comfortable articulating identity when it involves abstract categories or groups of people. At the level of daily lived experience, they tend to repress the articulation of difference—a practice that has historically served to maintain a sense of social harmony through the minimizing of attention to social difference. Indeed, in a society built around the idea of being ordinary, specifically assigning difference to particular individuals can be both disruptive and stigmatizing.

The resistance to articulating difference that I encountered in the Netherlands functions within social processes that work to reproduce Dutch social life within the context of the worldview of pillarization. These include local understandings of the self as produced in Dutch social networks and social silence not just around difference in general but around anomalies that people generally gloss as congenital (*aangeboren afwijkin-*

gen), which include genetic conditions. The Dutch tradition of tolerance and its system of pillarization provide a critical context for understanding how Dutch people deal with difference. It is within this context that I situate my subsequent discussions of the relationship between increasing scientific knowledge of genetics in the Netherlands and Dutch conceptions of identity. In chapter 3 I will elaborate how pillarization as a worldview informs and illuminates the ways Dutch people construe and manage genetic differences. Before considering the relationship between general Dutch attitudes toward difference on the one hand and Dutch medical, scientific, and social management of genetic difference on the other, I want to turn to an analysis of how specific attitudes toward genetics in the Netherlands are informed by the legacy of the Second World War.

REMEMBERING THE SECOND WORLD WAR

The legacy of the history of the Second World War for the Netherlands is important in my analysis for two primary reasons. First, widespread understandings of a shared experience of oppression during the war serve as an important reference for articulations of a distinctive Dutch identity. Second, attitudes toward genetics in the Netherlands are deeply informed by Dutch associations of genetic practices with Nazism. The Nazis' efforts to produce a "super race" through its eugenic program of racial hygiene have had a persistent influence on the interpretation and practice of contemporary genetics. An extended story of bicycles and bicycling highlights how an important symbol of national identity operates to help interpret the experiences of the Second World War.

Bicycling as a Symbol of Dutchness

Most mornings during my fieldwork in the Netherlands I joined the busy bicycle traffic as I rode my bike to the clinical genetics center where I carried out my ethnographic research. The traffic comprised people on their way to work, university students, and schoolchildren heading to their classes. Women and men carried as many as three children on their bicycles, taking them to child care or to school on their way to their own workplaces. One typically gray morning, upon arriving at the clinic, I waved to a clinic social worker who was just entering the building. After securing my bike along the canal I went inside and saw the social worker in the main office. She immediately remarked, "You are really Dutch now, with your bicycle!" and inquired if I had learned how to ride a bike since arriving in the Netherlands.

On another day, Joost, a senior physician, and I left the weekly prenatal diagnosis meeting of medical geneticists and obstetrician/gynecologists at the same time, chatting on our way out of the building. Joost expressed surprise when he saw me walk over to the overflowing bike stands near the building and begin unlocking my bike. He said, "Oh, you have a bike!" and I replied, "Yes, I love it." He, too, asked me if I had learned to ride in the Netherlands. I said no, that I had learned in the United States as a child, but he persisted in connecting the ability to ride a bike and Dutchness, saying, "Oh, you have a bit of the native blood in you." Then he looked at the bike to make sure it was a real Dutch bike. He expressed approval that the little front light did not work because Dutch people typically ride rather run-down bikes for daily use in the city so as not to attract bike thieves.

These kinds of encounters were not uncommon; people were often surprised that I knew how to ride a bicycle, and I was frequently asked if I had learned to ride a bike since arriving in the Netherlands. When I discussed this with a Dutch friend she jokingly suggested that I tell people I had learned to ride a bike in China because, according to her, bicycling was something Dutch people understood that they had in common with Chinese people. The United States is not thought of as a country of bike riders, whereas cycling is an integral aspect of the daily lives of most Dutch people both in rural and urban areas. People use their bicycles to run errands around town, to go grocery shopping, to go to work, and to take their children to child care or school. They also use them for leisure activity—riding out of town on long bike rides on sunny weekends or going for annual rides to see spring flowers blooming. The sociality of cycling is quite striking in the Netherlands. One frequently sees someone riding a bike with a second person perched sideways on the back of the bike. I often saw a couple that I guessed to be in their late sixties or early seventies ride by my house, he peddling along and she sitting sideways on the rack above the rear wheel.

Bicycles have also become something of a symbol of the problems associated with recent immigration. That Turkish and Moroccan immigrant women in the Netherlands at the time of my fieldwork often did not know how to ride bicycles was recognized as a significant social problem. This inability marked them as different and was seen by many as one of numerous obstacles to their integration into Dutch social life. Neighborhood centers throughout the country offered bicycle lessons especially tailored to these women.

Bicycles and the ability to ride one persist as important signifiers of Dutch identity. In a book on Dutch legal culture, for example, the authors emphasize Dutch values of egalitarianism by writing that "even the Queen occasionally rides a bicycle, demonstrating she is truly Dutch" (Blankenburg and Bruinsma 1994:75). In June 1997, during a meeting in Amsterdam focusing on European unification, particularly the problems associated with implementing a common currency, the mayor of Amsterdam gave a gift of a bicycle to each of the European leaders in attendance. Ironically, the German and French presidents, Helmut Kohl and Jacques Chirac, respectively, declined to ride (Whitney 1997:A3).

The story of bicycles and bicycling in the Netherlands highlights the unexpectedly deep significance contained in what many are bound to assume is a simple activity. Bicycling is also inextricably connected to national identity and memories of the Second World War. In his history of the destruction of Dutch Jewry during the war, J. Presser discusses the series of decrees implemented by the Nazi occupation forces that served to bar Jews from participating in public life. Access to a bicycle was (and for the most part continues to be today) a central component of access to the public sphere of social life. Presser writes,

> By Decree No. 58/1942 of May 21 Jews had previously been ordered to register their bicycles. The *Jewish Weekly* of June 12 informed its readers that the necessary forms were now available, how much one had to pay for the privilege of getting them, and that the completed forms had to be submitted before June 30. This was on the orders of the Reichskommissar himself (who promised that the bikes would not be confiscated). But meanwhile Rauter had concocted a regulation of his own, No. 14 of June 22, ordering Jews to hand in their bicycles not later than 1 p.m. on June 24, that is within forty-eight hours, in good working condition and "not forgetting spare tyres and tubes." To make quite sure, Article 2 added: "It is prohibited to sell hire or lend bicycles to Jews." (Presser 1965:129)

Toward the end of the war the Nazis began shifting many resources (including all types of vehicles supporting the German occupation) to Germany, where they were attempting to hold back Allied forces. As their means of transportation were shipped out of the Netherlands, German military personnel began commandeering Dutch citizens' bicycles. Today, at soccer games involving German teams or upon seeing German visitors fulfilling the stereotype of rude tourists, Dutch people will still occasionally say, "I want my bike back."

Remembering the War

The ongoing relevance of discussions about Dutch experiences and activities during the Second World War suggests the persisting significance the war has for Dutch people today. That significance is evidenced in the contemporary commemoration of those who died (*Dodenherdenking*) and the celebration of the nation's liberation (*Bevrijding*), observed annually on May fourth and May fifth, respectively. In the weeks leading up to these holidays the various television broadcasters run a series of programs related to the war and war experiences. May fourth involves a solemn commemoration for which people gather before eight in the evening in their town squares to lay wreaths for those who died in the war. The fifth of May, by contrast, is a jubilant event marked by celebrations and parades in recognition of liberation.

The war has been memorialized and mythologized in books, films, and television programs. A single history of the war by the Dutch historian Louis de Jong was, by 1987, a twenty-one-volume set nearly thirteen thousand pages in length (de Jong 1969–88). Shetter remarks on the popularity of this work by explaining that "it has achieved a widespread popularity not usually enjoyed by the work of professional historians, precisely because it . . . projected a picture of a crucial break in history from which moral lessons can be drawn. Many ordinary people buy and read around in the volumes. . . . As they appear they are reviewed in detail in the popular press. Most volumes appear on the non-fiction bestseller list, some in first place" (Shetter 1987:230). According to Shetter, by 1987 between three and four thousand books about the war, around 90 percent of them nonfiction, had been published in the Netherlands. Demand for material touching on war themes persists. Just after my arrival in the Netherlands one of the television broadcasters began rebroadcasting a popular mini-series titled *Zommer Van '45* (Summer of '45) based on the book *Children of the Liberation*. The program, in Dutch and English (with subtitles for the English), followed the stories of several young Dutch women and their encounters with the Canadian soldiers who made up part of the liberating forces. During my fieldwork, the year before the fiftieth anniversary of the liberation, one Dutch publisher put out a year-long series of newspapers called the *War News* (*Oorlogskrant*). The paper reviewed the war experience by reprinting, in each installment, the major news stories from fifty years before.

Additionally, the war is memorialized in monuments and plaques throughout the country that call attention to local sites of Nazi atrocities. Foremost among these is the house where Anne Frank, her family, and their friends hid from the Nazis for most of the war. Located on the Prinsengracht in Amsterdam, the Anne Frank House is one of the most frequented tourist sites in all of the Netherlands. It preserves intact the secret living quarters in which Frank wrote her famous diary (Frank 1952) about living in hiding during the Nazi occupation. The monument also houses a foundation dedicated to calling attention to contemporary fascist movements wherever they arise.

On a more modest level, I encountered a small but powerful reminder of the legacy of the war each day when I entered the genetics center where I was conducting some of my research. There by the door was a plaque that read:

> *"Het kind is niet meer."*
> *Gen. 37:30*
> *IN DIT GEBOUW WAS SINDS 1871 GEVESTIGD*
> *HET CENTRAAL ISRAELITICSH WEESHUIS.*
> *JOODSE WEESKINDEREN VONDEN HIER*
> *EEN VEILIG THUIS.*
> *IN 1942 WERDEN ZIJ, HUN BEGELEIDERS EN*
> *HUN DIRECTEUR, VAN HIER WEGGEVOERD.*
> *ZIJ KEERDEN NIMMER WEER.*
> *"MOGEN HUN ZIELEN GEBUNDELD WORDEN*
> *IN DE BUNDEL VAN HET EEUWIGE LEVEN."*
> *26 APRIL 1992 23 NISAN 5752*
> ["The child is no more."
> Gen. 37:30
> IN 1871 THE CENTRAL JEWISH ORPHANAGE WAS ESTABLISHED
> IN THIS BUILDING.
> JEWISH ORPHANS FOUND A SAFE HOME HERE.
> IN 1942 THEY, THEIR CAREGIVERS,
> AND THEIR DIRECTOR WERE TAKEN AWAY.
> THEY WERE NEVER SEEN AGAIN.
> "MAY THEIR SOULS LIVE FOREVER."
> 26 APRIL 1992 23 NISAN 5752]

Even though I was acutely aware of Dutch geneticists' highly self-conscious attempt to define their work against Nazi science, few of them seemed to note this ironic reminder of the legacy of the Holocaust.

Memories of the war are not only rooted in memorials and symbolic interpretations of war experiences. The war also deeply affected the nation at the material level. Shetter suggests that in comparison to other countries in Western Europe occupied by Nazi forces, the material effects of the occupation in the Netherlands were quite severe. Between invasion and liberation, Shetter reports, "although the population increased by about 5%, 10% of all housing was destroyed and there was virtually no building done. 30% of industry was destroyed and 40% of productive capacity lost. The transportation system was at a standstill, with only 20% of the rolling stock left. Most of the country's cattle and poultry had been taken. Rotterdam was the most heavily damaged Allied city of the War, and the Northeast Polder, the Wieringermeer Polder and the island of Walcheren in Zeeland were all flooded" (Shetter 1987:229–30). No discussion of Dutch experiences during the Second World War would be complete without mention of the *hongerwinter*, or hunger winter. Some of the most stirring memories of the war revolve around the experiences of the winter of 1944–45. That winter was unusually severe and food supplies were cut off, chiefly in the western part of the country near Rotterdam. As a result, many Dutch people starved just as the war was ending and much of Europe was already liberated (Shetter 1987:229).

It is not only people who themselves experienced the occupation who feel the importance of commemorating it. Each time the topic came up with young people they told me that they personally felt it important to celebrate the May fourth and fifth holidays. Rudie, an engineer in his early thirties, told me he could not imagine not going to his town square on the fourth of May to commemorate the dead because he felt that he owed those people something for his own life. He then talked a bit about the experiences of his parents and his wife's father during the war. Tineke sent me pictures of Canadian veterans marching through her town on the fiftieth anniversary of the May fifth liberation. She wrote, "I can't believe what these sweet old men did for us."

In his essay about Dutch character, van de Braak states that the German occupation during the Second World War left deep scars on Dutch society. He argues that the occupation has had a deeper influence on Dutch people than similar experiences had on the Belgian or French (van de Braak 1993).

Shetter argues that Dutch wartime mythology has centered on the theme of opposition to the occupier (Shetter 1987:230–33). In a lecture given to college and university students, the Dutch historian C. Klep (1994) explained that immediately after liberation the war was thought of in very "black and white" terms as a battle between good and bad. A relatively recent component of interest in the war, however, is a reevaluation of Dutch activity during the war (Hirschfeld 1988). This reevaluation involves questions about Dutch passivity and complicity during the occupation. According to Klep (1994), by the late 1960s people were beginning to interrogate the meaning of the apparently dominant strategy of accommodation during the occupation.

Van de Braak, for example, wonders how to make sense of the fact that a higher proportion of Jews were deported from the Netherlands during the war than from other countries, given the importance of the Dutch value of tolerance (van de Braak 1993:15). Indeed, questions about whether either the state or ordinary people could have done more to prevent the destruction of Dutch Jews are at the center of contemporary discussions about the war. These questions are so salient that they resurfaced in the wake of similar questions about the massacre of the Muslim population by Serb forces in Srebenice when the city was under the protection of Dutch military forces serving with the United Nations peacekeeping units in Bosnia (Simons 1998).

Genetics and the Legacy of the Second World War

In his comprehensive work on the history of eugenics, Daniel Kevles points out that the "specter of eugenics hovers over virtually all contemporary developments in human genetics" (Kevles 1995 [1985]:ix). He links popular negative attitudes toward eugenics to "the Nazi horrors [which] discredited eugenics as a social program" (1995 [1985]:xi). According to Kevles, "The barbarousness of Nazi policies eventually provoked a powerful anti-eugenic reaction" (1995 [1985]:118). Central to Kevles's analysis is the way biologists in the United States and Britain came together after the Second World War to fight to establish human genetics as a scientific discipline focused on disease and clearly distinguished from eugenic biases and policies (Kevles 1995 [1985]:ix; see also Keller 1992).

For many Dutch people Nazi science and the legacy of the Second World War represent the very antithesis of the worldview of pillarization. Dutch people are highly attuned to the history of Nazi intolerance toward difference as manifested in its attempt to produce a super race through its eugenics program of racial hygiene (Proctor 1988). Dutch people often employ explicit references to these polar extremes of tolerance and intolerance to help them make sense of the relationships among genetics, identity, and belonging in Dutch society.

The Nazi program of racial hygiene which incorporated strong eugenic policies was, contrary to what many believe, not an example of bad science or science gone awry. One of the most profound insights Robert Proctor (1988; 1995) develops in his analysis of racial hygiene is that the program was one through which widely accepted scientific theories developed by major figures in biology were enacted into social policy. Indeed, both Kevles and Proctor convincingly demonstrate that Nazi eugenic policies were based on arguments published by prominent biologists in the United States and England. In his article examining the phenomenon of the Nazi destruction of "lives not worth living," Proctor illustrates how physicians were central to constructing the mentally ill and those designated as handicapped, along with Jews, as threats to public health. Proctor demonstrates how policies intended to exterminate large groups of people flowed logically from the way they were defined as deficient.

Nazi intolerance of difference, as exemplified in the policies of racial hygiene, is a symbol against which Dutch people distinguish themselves from their German neighbors. In his lecture, Klep (1994) drew on Dutch values about tolerance to explain the failure of Nazi occupiers to fully "Nazify" the Netherlands. He argued that German occupiers were "ideologically and politically unsuccessful in Nazifying the Netherlands . . . because of pillarization and tolerance." He explained that "tolerance was a basic attitude. Not among fifty to sixty percent of the population, but basic [to the society as a whole]."

Invasion and Occupation

Interpretations of what happened in the Netherlands during the Second World War are hotly debated and intensely political. Although the Netherlands remained neutral during the First World War, Hitler invaded the country early in the Second World War. Klep, a specialist in military history, outlined the invasion in his lecture to the American students by discussing Dutch mobilization against it. He described the Dutch army as poorly armed and ill-prepared for the German military assault. The Dutch military, for example, had no tanks and few trucks. They relied on their ability to flood a large portion of the country through their extensive water management system but were inadequately prepared when the Germans flew over the floods to drop troops inside Dutch fortifications. Klep described the Dutch military as a "bicycle army."

As Klep put it, Hitler intended to take the Netherlands in two days, but the Dutch "held out for four days." The Dutch surrendered on May 14, 1940, and remained an occupied nation until the country was liberated on May 5, 1945. Although the Nazis did not succeed in Nazifying the Dutch population, Klep argued that they did force accommodation, a strategy that was adopted by fully 90 percent of the population. A strategy of accommodation, according to Klep, entailed turning inward to focus on family life, using "families as a cocoon," and trying to continue living life as normally as possible without interacting with the German occupiers.

Recent interpretations of the war point to the fine line between accommodation and either resistance or collaboration. Klep made the point, for example, that as the war progressed, accommodation became increasingly difficult. Ration cards had to be obtained from German offices, and Dutch people were required to carry identity cards that had to be shown at every roadblock. In addition, Dutch products were shipped to Germany to supply Germans and the German military. Klep concluded that under occupation Dutch social life became "more and more German."

The wartime requirement of carrying an identity card has important consequences for contemporary interpretations of the idea of a genetic passport. The concept of the genetic passport gained popular attention after the first broadcast of Wim Kayzer's four-part television program *Beter Dan God* in 1987 and persisted well into the time of my fieldwork in 1994. Until 1994, when a policy requiring identity cards was enacted, the only time in modern Dutch history that Dutch citizens had been required to carry identity cards was during the Nazi occupation. Thus, the possibility of introducing something like a genetic passport inevitably invokes the specter of Nazi oppression and intolerance.

A number of the people interviewed in *Beter Dan God* made direct links between contemporary genetics, Nazi science, and the idea of a genetic passport. In the second installment of the program the commentator quotes a series of individuals regarding issues related to genetic testing. He quotes the molecular biologist Huub Schellekens, who explicitly introduces Nazi eugenic practices: "The methodology is different. Hitler murdered [*vermoorden*] people or they were forced [*gedwongen*] to be sterilized. Now it happens in a roundabout way [*omweg*], through ascertaining [*vaststellen*] an abnormality in the uterus and abortion. The result and the basis for it is the same" (Kayzer 1987:2). Later in the same installment the commentator discusses the concept of a genetic passport in relation to the Nazi deportations of Jews and other Dutch people:

> Almost everyone we spoke with had no doubt. The genetic passport will come. At the same time no one is at ease [*rustig*] about the safety of those data. "The government should get involved [*ingrijpen*]," says a lecturer in biochemistry, "but the same government allows it to happen already and, moreover, the same government has an immense interest in genetic information. I hope not to live in such a society. It is a dreadful [*verschrikkelijk*] picture of the future." The director of the clinical genetics center in Utrecht urges that the medical records of patients should not be removed after ten years as they are now, but should be kept much longer and preserved for genetic research [*erfelijkheidsonderzoek*]. In the meantime, we filmed at the public registry [*bevolkingsregister*] in the city of Utrecht. The written data files have been copied onto computers. Meanwhile, nobody can truly explain why this arrangement is really necessary. How recently was it that 140,000 Dutch people were deported and killed? How recently were the public registry pages being exploded [by the resistance] to stop those deportations?

Today one of the most powerful symbols of the Nazi program of racial hygiene and German intolerance in the Netherlands is the deportation of Dutch Jews during the Second World War. Like the Jewish populations of many European nations, the Jewish community of the Netherlands was decimated in the war. According to one source, of the 140,000 people registered as "full-Jews" under German racial categories in the Netherlands in 1941, only 5,450 returned to the Netherlands after the war (Presser 1968:538–39). Including the 140,000 Jews, 200,000 Dutch citizens lost their lives in the war. Furthermore, 400,000 non-Jewish men were deported to Germany for forced labor (Shetter 1987:229–30).

The deportation and death of so many thousands of Dutch Jews continually raise questions about Dutch complicity during the war and, therefore, about the gap between socially valued ideals and the reality of everyday life. It is in this context that virtually all activities associated with genetics in the Netherlands come to be defined in opposition to the values of Nazi science. Questions about Dutch complicity in the deportation of Jews during the war remain salient in Dutch society in large part because of the challenge they represent to the Dutch ideal of tolerance. These questions keep Nazi science and racial hygiene in the foreground of discussions about the ethics of medical and scientific practices of genetics in the Netherlands. In *Beter Dan God* this connection is made explicit in a comment made by Bernd Clees, a professor of social and workers' rights who served as an advisor to the European Parliament about new genetic technologies:

> The consequences, if we let them occur—alas, they already have—are also that people can be forced to come to the Central Labor Exchange for professional advice on the grounds of a genetic disposition. A "genetic profile" would then be decisive. Moreover, with the shrinking of the employment market, I see the danger that through developments in pre- and post-natal diagnosis, groups of people will be selected, corresponding with the specific ideas concerning "social hygiene," that we know from the past. Where this manner of thinking can degenerate can be summed up in one word: Auschwitz. We surely will do it with much less blood, with as clean a solution as possible—sterilization, or, what the following step is, eventually [genetic] manipulation.

This is an example of how people weave references to the war and Nazi policies into critiques of contemporary genetics. In my interviews people frequently made direct references to Nazi science in discussing genetics. The vast majority raised this issue by referring to the Nazi goal of producing a super race. They often did so by making distinctions between using such technologies for medical purposes, which they generally supported, and using them for the purpose of producing *een supermens* (a super race), a use they unequivocally opposed.

Contemporary questions about the legacy of the Second World War and Dutch complicity in the deportation of Dutch Jews challenge ongoing concerns central to Dutch identity having to do with the ideal of tolerance. In the following chapters I illustrate how such questions shape Dutch attitudes toward genetics, genetic practices, and differences associated with genetics. I do so, particularly in chapter 5, by examining how people deal with the significance of the Second World War in daily practice as they seek to engage genetic knowledge and practice on the one hand and to critique the potential consequences of that knowledge and practice on the other. I will argue that such distinctions, which rest on Dutch memories and experiences of the Second World War, ultimately distinguish between modernity and modernity gone awry.

Chapter Two

GENETICS AND THE ORGANIZATION OF GENETIC PRACTICE IN THE NETHERLANDS

The institutional arrangements within which human genetics research and clinical practices occur are highly structured in genetics centers in the Netherlands.[1] The genetics centers are the primary resource for any individual seeking medical information about genetics or genetic counseling. The Dutch system of health insurance allows all legal residents to take advantage of these services. The centers serve as sites for encounters between medical professionals, researchers, and the general population as they mutually negotiate the meaning of genetic knowledge in the Netherlands. I argue that the centralized organization and production of medical and scientific knowledge about genetics in the Netherlands has two significant consequences. First, it builds a coherent and unified profession of geneticists throughout the country. Second, it thereby enhances the production of an authoritative discourse about the nature and meaning of genetic knowledge.

The organization of genetic practices in the Netherlands constitutes a powerful institutional framework through which Dutch people encounter authoritative knowledge about genetics. I highlight here the characteristically Dutch organization of a major medical and scientific endeavor while illuminating the myriad social relationships involved in producing and managing scientific knowledge about the body in the Dutch practice of medical genetics. In so doing, I want to emphasize how little determinism there may be in genetics and how much force is exerted by profound and multiple forms of complexity—for example, the physiological complexity of different bodies or the multicausal complexity of a particular condition that are too often collapsed in the shorthand term *genetic*.

HUMAN GENETICS AND THE PROCESS OF GENETIC INHERITANCE

Human genetics involves the study of human biological inheritance. Contemporary work in the field is based on a combination of knowledge developed over many years, ranging from Gregor Mendel's description of recessive and dominant inheritance patterns in peas (Mendel 1865/1866) up to recent and rapid developments in molecular biology. In his paper Mendel proposed the idea of the existence of discrete units of biological inheritance—which today are called genes—and suggested that pairs of these independent factors, one inherited from each parent, determine simple genetic traits, including some genetic disorders. By 1900 scientists had described chromosomes, molecular structures that also showed specific patterns of inheritance—patterns corresponding with the inheritance patterns of genes.

A human genome is the sum of the estimated three billion pairs of bases—adenine, guanine, thymine, and cytosine—that form the approximately twenty-eight thousand to thirty-five thousand genes in humans. These genes are widely considered to determine the expression of all genetically linked human traits. The double helix, described by James Watson and Francis Crick in 1953, defines the structure of this genetic material—deoxyribonucleic acid (DNA) (Watson and Crick 1953). The double helical structure of DNA is itself twisted up into the forty-six chromosomes found in most human cells. Each person has his or her own, unique (unless he or she is an identical twin) genome.

Research in human genetics has accelerated since the late 1980s, when the governments of the European Community, the United States, and Japan committed themselves to funding the Human Genome Project (HGP). The project aimed to map the estimated three billion base pairs constituting the human genome. Many scientists expect knowledge produced through the HGP to dramatically change the nature of biological sciences, medical treatment, and humans' understandings of themselves in terms of their biological functionings. Through procedures involving recombinant DNA techniques, in vitro and in vivo procedures, and genetic counseling, the research makes real the possibility of isolating genes for particular traits and of preventing the expression of specific traits.

The molecular biologist and Nobel laureate Walter Gilbert calls the human genome the Holy Grail of biology (Hall 1990), while others describe it as "the key to what makes us human" (Kevles and Hood 1992:overleaf). DNA itself has come to be known as the "mother molecule of life" (Keller 1983:5), and contemporary research in genetics has been described as "the most astonishing scientific adventure of our time" (Bishop and Waldholz 1990). These descriptions, resonating with deeply held beliefs about religion, truth, science, and human life, reveal that scientists attach profound significance to the material encoded on the double helical structure of DNA.

Gilbert has written that with the mapping of the human genome "a whole variety of human susceptibilities will be recognized as having genetic origins" (Gilbert 1992:94). He believes that as scientists increasingly understand the functioning of human genes, they will come to see that individual limitations are inscribed in genetic codes (Gilbert 1992:96). Gilbert suggests that eventually people will be able to carry a compact disc with their genome encoded on it, and he makes this point in lectures by pulling out a compact disc and telling members of his audience, "This is you" (Dreyfuss and Nelkin 1992:319).

Early results of increasing knowledge about genetics foster the extension of such existing services as genetic counseling, risk assessment, and carrier and prenatal screening (Cowan 1992:245). An explicit component of human genome research is the idea that scientists eventually will find molecular treatments for genetic disorders (Cantor 1992:94–95; Hood 1992:159–60; Wexler 1992:240–42). The treatments theoretically possible for such disorders include cell therapy; chemical inductions at specific loci of the genome designed to produce desired traits; specifically targeted pharmaceuticals; transduction into humans of genetic information incorporated into microorganisms; and the use of artificially manufactured genes that might be inserted to replace or override genes producing undesired or debilitating traits.

The deterministic understanding of genes described above represents the dominant central dogma (Keller 1983) or the prevailing paradigm (Kuhn 1962), in both scientific and popular conceptualizations of genetics, until about 2000. In 1957 Crick emphasized the determinism at the heart of what he himself called the central dogma of biology, stating, "Once 'information' has passed into protein *it cannot get out again.* In more detail, the transfer of information from nucleic acid to nucleic acid, or from nucleic acid to protein may be possible, but transfer from protein to protein, or from protein to nucleic acid is impossible" (Crick 1957:153). Building on Crick's assertions, the molecular biologist Jacques Monod also displayed a highly deterministic understanding of genes, saying that "what molecular biology has done, you see, is to prove beyond any doubt . . . the complete independence of the genetic information from events occurring outside or even inside the cell—to prove by the very structure of the genetic code and the way it is transcribed that no information from outside, of any kind, can ever penetrate the inheritable genetic message" (quoted in Judson 1979:217). This central dogma has proven quite tenacious and continues to maintain a central place in scientific research today.

There is debate, nevertheless, within the scientific community about just how determining genes actually can be. The biologist Ruth Hubbard and her coauthor Elijah Wald (1993), for example, argue that the metabolic pathways occurring between a gene and its expression are so numerous and complex as to call into question whether one can have a gene "for" a particular molecule, trait, or disease. They assert that "when scientists talk about genes 'for' this or that molecule, trait, or disease they are being fanciful. They attribute excessive control and power to genes and DNA, rather than seeing them as part of the overall functioning of cells and organisms" (Hubbard and Wald 1993:53). For Hubbard and Wald, "Inherited factors can have an impact on our health, but their effects are embedded in a network of biological and ecological relationships" (Hubbard and Wald 1993:63). They argue that such relationships, involving genes but not determined by them, help explain why the same gene can lead to different consequences for different people (Hubbard and Wald 1993:64).

In his work on biology and ideology the geneticist Richard Lewontin elaborates on what he considers the ideological basis for many scientists' faith in the human genome as the source for explicating the essence of humanity. He, like Hubbard and Wald, stresses that "we are not determined by our genes, although we are surely influenced by them. Development depends not only on the materials that have been inherited from parents—that is, the genes *and other materials* in the sperm and egg—but also on the particular temperature, humidity, nutrition, smells, sights, and sounds (including what we call education) that impinge on the developing organism" (Lewontin 1991:26). Events leading up to and following the announcement of the completion of a first rough draft of the human genome in 2000 emphasize the complexity the project's critics had long been pointing to. The year before, Jesse Gelsinger, a teenager participating in a gene therapy trial at the University of Pennsylvania, died as a consequence of his therapy. In the media attention that his death garnered the public learned that not only had others died in gene therapy trials, but also that every gene therapy trial had (and has since) failed (Lewontin 2001; Stolberg 1999). This news highlighted the fact that, in spite of its mapping successes, the hundreds of millions of dollars in public funds spent on human genome research had not yet translated into the kinds of significant high-tech medical interventions the project's promoters had long been promising. Keller, who has described the twentieth century as the century of the gene, suggests that rather than finding the "secret of life" at the level of the genome, what has been found is life's enormous complexity (Keller 2000:7–8). It is to this complexity that researchers have turned in the postmapping phase of human genome research.

In fact, as Lewontin (2000) has argued, all thoughtful scientists (and, in fact, many nonscientists) understand that genes are only a part of any story about embodied conditions. While many scientists focused on the centrality of genes and suggested a future in which the mapping of the human genome would lead directly to treatment for genetic diseases, critics like Hubbard, Wald, and Lewontin stressed the social relations and priorities underlying scientists' focus on genes. Today, there continue to be reports of genes for extraordinarily complex embodied conditions, especially coming out of the work of researchers in the fields of evolutionary psychology and sociobiology. Indeed the past decade has witnessed claims about genes for "thrill seeking," criminality, homosexuality, and even for political views and religious beliefs (Carson and Rothstein 2002; Parens, Chapman, and Press 2005). By 2000, however, genomic scientists had widely abandoned such a simplistic perspective in favor of a focus on complexity. In the aftermath of the completed mapping of the human genome, these researchers were articulating the need to develop research strategies that would lead to better understandings of gene-gene interaction and of genes in relation to their environments (Taussig 2005). The images, discussions, and debates concerning DNA, the double helix, and genetics reflect a wide range of powerful beliefs about individuals, society, nature, biology, biological inheritance, identity, and kinship.

Genes and the chromosomes upon which they are located are, as noted, made up of pairs of bases—adenine, guanine, thymine, and cytosine—known as deoxyribonucleic acid (DNA) (Watson and Crick 1953).[2] The double helical structure of DNA is twisted up into the chromosomes found within the nucleus of every cell of an individual's body. Typically, humans have forty-six chromosomes located within the nucleus of each of the cells in their bodies, except their reproductive cells. Human reproductive cells —gametes known as sperm and ova—usually contain twenty-three chromosomes each. At conception these two sets of chromosomes come together to create a zygote with forty-six chromosomes. The zygote undergoes rapid cell division, with the potential to become a newborn with over a trillion cells. Geneticists pair the forty-six human chromosomes in twenty-three sets, which they identify by number and letter in visual representations of chromosomes known as karyotypes. Forty-four of these chromosomes are called autosomes, and two are called sex chromosomes, designated as X and Y chromosomes because of their respective shapes. Geneticists describe a normal female karyotype as 46,XX and a normal male karyotype as 46,XY.[3]

Prevailing medical and scientific understandings of inheritance patterns are integral to contemporary genetics practices. At the core of these understandings is the idea, based on Mendel's and others' subsequent work, that the biological inheritance of certain traits occurs in predictable patterns. These include recessive, dominant, and X-linked patterns of genetic inheritance. The rules that govern these patterns of inheritance are based on the understanding that for every trait determined by a single gene, an individual actually has two genes—one from each parent. If a trait is recessive, an individual must have two copies of the gene to actually exhibit that trait. If an individual has only a single copy of the gene, she or he is considered a carrier of the gene but does not exhibit the trait. Examples of recessive disorders include cystic fibrosis and sickle cell anemia. A trait is considered dominant if a single copy of the gene for that trait is enough to ensure that an individual will display that trait. Some forms of muscular dystrophy are dominant, as is the neurological disorder known as Huntington's chorea. If a trait is X-linked it means that it is a recessive trait for which the gene is located on the X chromosome. Because females have two X chromosomes they are usually unaffected by X-linked traits because they have a second copy of the gene on their other X chromosome. Because males have an X and a Y chromosome, however, if they inherit a gene for an X-linked trait they will exhibit that trait since they have no compensating gene on their Y chromosome. Hemophilia and colorblindness are common examples of X-linked disorders. These patterns of inheritance, known as simple Mendelian patterns of inheritance, highlight the distinction between phenotype and genotype. Phenotype refers to genetic traits an individual exhibits, while genotype refers to the genes in an individual's cells whether or not the traits associated with the genes are expressed in a particular individual.

Simple Mendelian patterns of inheritance operate only in specific cases, usually those associated with a single gene or genes that work together in predictable ways. Geneticists do not have as clear an understanding of the inheritance patterns of traits determined by multiple genes—those that are polygenic. Geneticists are also aware that most traits are multifactorial, in that they may have a genetic component but ultimately are the result of a combination of factors, including environment.

Mutations at the level of DNA, such as those described above, involve the absence, known as a deletion, of one or more base pairs or the presence of too many base pairs within a gene. In addition to genetic disorders connected to specific genes at the level of DNA, there are genetic disorders associated with chromosome anomalies, which involve variation in the number or structure of chromosomes in an individual's cells. One of the most common chromosome anomalies involves the presence of an extra chromosome in addition to the forty-six most humans have. Such a condition is labeled a trisomy because instead of the usual two, an individual has three of a particular chromosome in her cells. The most commonly known trisomy is trisomy 21, known widely as Down syndrome. Variations in the expected number or structure of sex chromosomes also occur. Such variations are labeled specifically as sex chromosome abnormalities. Individuals who are biologically female tend to have two X chromosomes, one from each parent, while those who are biologically male tend to have an X and a Y chromosome, the X from their mother and the Y from their father. Variations in the expected number of sex chromosomes occur frequently enough that some are named as syndromes, such as Turner syndrome (in which individuals have a single X chromosome and no complementary X or Y and described as 45,X) and Klinefelter syndrome (most often characterized by the presence of an extra sex chromosome, so that individuals with the syndrome are usually described as 47,XXY). In some cases, chromosome analysis reveals variation in the expected structure of chromosomes. These cases can involve a missing or an extra piece of a particular chromosome, or what is known as a translocation—a situation in which pieces from two or more chromosomes break off and reattach in new combinations.

THE ORGANIZATION OF GENETIC PRACTICES

The burgeoning genetics practices I encountered in the Netherlands are a relatively new phenomenon, organized in their present form only since 1977. There are eight genetics centers in the country devoted to researching and diagnosing genetic anomalies in humans on both a pre- and postnatal basis. Each of the centers is affiliated with, or housed within, one of the eight teaching hospitals in the Netherlands. These centers are highly efficient and integrated both horizontally and vertically on a national level. They are vertically integrated in that each genetics center houses a clinical practice, a department of social workers or psychiatrists specializing in problems associated with genetic disorders, both DNA and cytogenetic diagnostic laboratories, and research laboratories. The clinics play a central role in the processes of normalization produced through genetic practices (I elaborate on these processes in chapter 3). In what follows I examine the organization and operation of the institutional structures that enable the systematic, widespread instantiation of that normalization throughout Dutch society.

Genetic practice in the Netherlands is horizontally integrated through regular national meetings of the various divisions making up each center. For example, on the third Thursday of each month the clinical geneticists from all eight genetics centers come together for training, for presentations about developments in their field, and to discuss difficult cases. The morning is devoted to training sessions for residents from all of the centers and to a meeting of the heads of the eight clinical practices. The clinicians spend the afternoon in a meeting, discussing new developments in the field and at the various centers as well as presenting and discussing complicated cases. Every two months the geneticists working in DNA diagnostics convene, and twice a year the staff working on psychosocial aspects of genetics come together. Several times a year the clinicians and geneticists working in both the cytogenetic and DNA diagnostic labs meet on a national level; twice a year there is a meeting of the Dutch Society of Human Geneticists, at which the staff from all divisions of all of the centers come together for the presentation of scholarly papers.

In addition to the integration of departments within and across each of the centers, the centers maintain close relationships with various other departments at the teaching and other hospitals, such as obstetrics and gynecology (ob/gyn), pediatrics, neurology, and metabolism. The physicians and technicians staffing ob/gyn departments at the teaching hospitals, for example, are actively involved in prenatal diagnosis; they counsel pregnant women about the possibility of prenatal testing and conduct the amniocenteses and chorion villus sampling that provide the cells necessary for prenatal diagnosis of genetic anomalies. The highly integrated organization of the centers gives the clinical and research geneticists who work in them an enormous level of control over the practice of genetics in the Netherlands. The eight centers are the only places in the country where genetic services such as genetic counseling or the clinical and laboratory investigation of suspected genetic disorders are available.

The institutional organization of genetic practices in the Netherlands serves the purpose of profession building in three ways. First, through such structured exchanges of information as the myriad regular meetings, the centers actively foster bonds among senior genetic practitioners while initiating junior clinicians into the norms and vocabulary of the profession. Second, through outreach programs targeted at the medical profession beyond the centers, such as satellite clinics and the education of general practitioners, geneticists work to instantiate scientific and medical knowledge about genetics throughout the medical profession. Third, through outreach programs aimed at the general public, ranging from daily encounters with patients to appearances in major media, geneticists affect popular understandings of and discourse about genetics. Together these profession-building activities powerfully shape the production, interpretation, and consumption of genetics in the Netherlands. They establish contemporary medical and scientific understandings of genetics as the dominant paradigm with which all other conceptions of genetics are invariably forced to contend.

A number of physicians discussed with me the value of the integration of the centers and their regular meetings. Though the issue of how the centers are organized came up in response to my direct questions about it, it also came up in other contexts. For example, in responding to my question about dealing with the enormous volume of information now being generated in the field of genetics, a clinic director spoke about the structure of the centers:

> **Hans:** Hmmm. How to deal with it? Keep working in a center so you hear from colleagues different things. And I think that is one of the most important things. Unlike in the U.S. I think where you can work as a clinical geneticist on your own or with two [people] I think it is a very

good system in Holland that you are obliged in fact to cooperate and to work with others and to listen to others; the only way I can keep up with literature is to listen to my colleagues on the many occasions that we talk to each other. . . .

KST: I have noticed that the centers are highly integrated, and I was hoping you could tell me what you think the positive and negative aspects of this integration are.

Hans: To start with the positive things . . . the fact that there are only eight centers in the Netherlands and . . . all genetic work in the Netherlands can only start with permission. So you have to have permission for everything, for genetic counseling, for DNA [analysis], for post-natal, pre-natal, biochemical analysis, so not our every clever boy or girl with a little bit of money can say 'okay I'm going to screen for metabolic disorder' no, you have to ask for permission and to fulfill certain terms, and the permissions are only granted to the centers. So an undesirable development is stopped by that. I think that is good so you can guarantee and look after quality and that is a very important thing also.

Another clinic director, responding to the same question about the positive and negative aspects of this integrated structure, told me that there were "mostly positives" and did not mention anything that he considered negative in his further discussion. What he did stress was that he believes the vertical and horizontal integration of the centers is a reflection of the fact that the Netherlands is very much a "consensus society," that the need for consensus is very strong, and the structure of the centers helps facilitate consensus about how to manage human genetics in the Netherlands.

In addition to integration at the national level, the various divisions of the centers are integrated into international aspects of medical and scientific practices. Many of the technologies used in the clinics and the laboratories were developed outside of the country and imported for use in the Netherlands. Staff from each of the divisions regularly attend various international scholarly conferences and read and publish in major international scientific and medical journals. This international integration is further highlighted by the pervasive presence of the English language in Dutch genetic practices. Much of the daily work done by Dutch geneticists is conducted in English. The bulk of the international journals they read and publish in are in English, as are various important computer databases and online services.

One clinic director considers English-language ability so essential that he formalized it into the training of clinicians. At one of the regular weekly meetings at this clinic, physicians are expected (and the residents required) to present their cases in English. One of the few rules the director of this clinic imposed on me was identical to one that had been placed on an Indonesian clinician who had visited the clinic prior to my arrival, namely, that during the segment of the meeting when the Dutch geneticists were speaking only English, I speak only Dutch. In this setting I was also occasionally called upon to referee vocabulary, proper usage, or pronunciation. One resident emphasized the importance of English for the geneticists' professional life. In discussing the significance of publishing in international journals she told me, "You always want to publish in an international journal. We think that if something is published in our Dutch journal, it is not very good because it was not good enough to be published in an international journal."

The bulk of my fieldwork in the centers focused on the clinical practice of genetics. My descriptions of the various departments within the genetics centers also reflect this focus. The clinic is a rich site for understanding the dynamic processes through which scientific and medical knowledge is produced, interpreted, and consumed because it is a significant location for encounters between basic science, medical practitioners, and people affected by medical problems and their supporters. As I demonstrate in the following chapters, the clinic is also a critical site for the production of bodily and social identities. The clinic is a place where scientists carrying out basic research on a particular gene confront people adversely affected by that gene. Conversely, the clinic is where people affected by genetic conditions and their supporters come into contact with powerful scientific and medical explanations of those conditions. These activities make the clinic a central site for the flow of information among different social domains.

CLINICAL GENETICS IN THE NETHERLANDS

The traditional goal of clinical work is to transform medical complexities into solvable, treatable, and curable problems. These clinical goals are often complicated in the case of genetics, where the availability of treatment varies widely depending on the type of disorder. In the case of Marfan syndrome, for example, beta blockers and cardiac surgery offer positive benefits in terms of quality and extension of life (Heath 1998).[4] Treatments for many genetic disorders, however, offer little relief to those affected, and in some cases there are no treatments available at all. An unexpected aspect of the clinical practice of genetics in the Netherlands is that it is an extremely expensive, well-elaborated, and rapidly expanding clinical practice that does not treat patients. If palliative treatments are available, general practitioners outside of the centers administer them. Clinical practices in the genetics centers focus on diagnosing and counseling people who have or are at risk of having genetic disorders in their families. Diagnosis typically marks the conclusion to a clinical interaction at the genetics centers. Because it only rarely results in a cure, physicians understand a genetic diagnosis as something that will persist for a lifetime and perhaps be passed to future generations. In this context, a diagnosis implicitly comes to inform a person's entire life span and beyond. As physicians communicate this understanding to their patients, the diagnosis becomes a powerful, enduring presence in their lives.

An important feature of the organization of health care in the Netherlands is that general practitioners (*huisarts*) make up the vast majority of physicians in the country—over 90 percent. All residents of the country are expected to have a general practitioner who refers them for most specialized care.[5] Medical genetics is an unusual exception to this rule in that gaining access to the clinics does not require referral from a general practitioner. The guidelines regarding clinical practices in the eight genetics centers are such that individuals are able to contact the centers themselves to speak with a medical geneticist and arrange an appointment at the center if the clinician deems such an appointment appropriate. Clinicians justified this arrangement two ways: First, they explained, not all general practitioners understand genetic disorders well enough to reliably refer eligible patients for the services the centers offer; second, clinicians also said that the state, in the interest of public health, has an interest in ensuring that as many of those who are eligible for genetic services take advantage of them. Knowing about many genetic conditions appears to aid public health because such knowledge can be and often is used in genetic counseling and prenatal testing to reduce the incidence of the birth of more people with those genetic conditions. This social benefit is also what justifies the expense of maintaining the extensive program in genetic research and services I describe in this chapter.

All services at the center for which an individual is eligible are paid for by the complicated system of private and public health insurances that cover all legal residents of the country.[6] These services include diagnosis of individuals with a suspected genetic disorder as well as prenatal and pre-pregnancy counseling. Should the counseling interview indicate that an individual is at increased risk of having a child with a genetic anomaly she is eligible for prenatal testing. At the time of my fieldwork, indications for prenatal diagnosis were as follows:

— Maternal age of thirty-six or older
— A previous child with a chromosome anomaly
— One of the parents is a carrier of a chromosome anomaly
— The mother is a carrier of an X-linked illness
— There is an increased risk of a child with a neural tube defect
— There is an increased risk of a child with a genetically inherited illness for which DNA diagnosis is possible
— There is an increased risk of a child with a metabolic illness for which enzyme diagnosis is possible
— A structural defect is observed in ultrasound
(de Pater and Bos 1993:51)

Anyone who did not fit one of these categories was formally ineligible for these kinds of services even if they were willing to pay out of pocket for them. In fact, there were occasionally cases of people not fitting these categories who were made eligible for prenatal testing because of a psychosocial indication, an example being a case in which a clinic social worker found that a pregnant woman's intense anxiety over her belief that there was something wrong with her fetus might affect the pregnancy and therefore recommended prenatal testing based on a psychosocial indication. In this case there was indeed a problem with the fetus, and in explaining the case to me the social worker pointed out that "she [the mother] was right."

The Physicians

In the United States, genetic counseling usually is conducted by individuals who are not physicians but who hold specialized degrees in genetic counseling (Rapp 1999); at the time of my fieldwork in the Netherlands, by contrast, all genetic counseling was conducted by the physicians who staff the clinical practices within the genetics centers. Some of these physicians were doctors with established specialties in medical genetics, while others were young physicians doing residencies in the discipline. One senior clinician told me several times that, given the rapidly expanding demand for genetic counseling and the increasing focus on human genetics in general, he did not think that all genetic counseling could continue to be done by physicians. He suggested that in the not-too-distant future routine genetic counseling would be taken over by nonphysicians trained in genetic counseling, as in the United States, while physicians with specialties in genetics would focus on the more complicated types of cases.

As in the United States (Rapp 1991), in the Netherlands the field of genetics, particularly but not only at the clinical level, is a highly feminized field. At the time of my fieldwork the majority of physicians in the clinic where I conducted much of my research were female. At that time, two of the senior clinicians were male and two female, one of the men being the director of the center. Of the eight physicians doing residencies in clinical genetics, six were female and two male. The physician specializing in cytogenetic diagnostics was male.

For several reasons there is a range of expertise among the senior physicians working at the clinics. First, there are differences in terms of the amount of time they have spent working in the discipline. Some have just recently finished their residencies while others have been working in the field since before the organization of the centers in 1977. Second, differences in seniority and experience mean the physicians may approach the discipline from different starting points. The current residency in medical genetics is a product of the contemporary organization of the centers and therefore has been fairly recently established. The most senior physicians, then, did not do residencies in medical genetics but came to the field on a variety of paths—through specialties such as pediatrics, neurology, or metabolism. Third, many of the senior physicians have subspecialties within medical genetics. Some of these specialties are recently developed or newly developing interests, while others reflect the prior foci that drew them to genetics in the first place.

There are also distinctions among the residents working in the various clinics. The residency for medical genetics in the Netherlands lasts four years, so these junior physicians are at various stages of their residency. Furthermore, although their training requires that the residents be exposed to all types of genetic disorders, they have special interests and tend to increasingly focus on these interests as they move through the levels of training.

In the chapters that follow I leave aside most of these types of distinctions in order to maintain the confidentiality promised to all informants. As a means of calling attention to an individual informant's general position within the centers while maintaining confidentiality, in most examples I identify individual physicians simply as senior or junior. Senior physicians are those who already have established specialties in medical genetics, while the junior physicians are those doing residencies.

Paths to the Clinics

People come to the clinical genetics centers in the Netherlands via a number of paths. Although many of the people coming through the clinics are referred by their general practitioners, people can and do call the clinics directly to arrange appointments. The Dutch genetics centers deal with an extraordinarily wide range of cases that fall into two broad groups: pre- and postnatal. My explanation of these types of cases illustrates the primary routes by which people come to the clinics.

Prenatal Cases

The prenatal cases the clinics deal with involve genetic counseling for people at risk of having a child with a genetic condition either because they have already had a child with such an anomaly or because they have a relative affected by a genetic condition. The style of counseling the physicians provide is known as nondirective counseling. Examples of prenatal counseling cases that occurred during my fieldwork at the center include Henk and Anneke, who were referred by their ob/gyn to see a clinical geneticist during their second pregnancy because it was discovered during a routine prenatal appointment that there was a history of a sex-chromosome anomaly in Anneke's family. In another case, Paul and Regina sought out advice directly from the clinic. They came because they are first cousins with a family history of Down syndrome, cystic fibrosis, and eczema, all anomalies medically defined as rooted in genetics; they wanted to investigate what they could expect if they had children. A third case involved Peter and Marianne. Marianne was pregnant through in vitro fertilization, and the couple sought prenatal diagnosis because Marianne has myotonic dystrophy, and they had decided they wanted a child only if it did not carry the gene for that dominantly inherited genetic muscular disorder.[7] The intention of these prenatal consultations is to give people the opportunity to evaluate the genetic consequences of their reproduction. In discussing the clinical practice of genetics with me, one clinician returned again and again to this theme of understanding the consequences of one's reproduction. She said repeatedly that the most unfortunate situation is for someone to have a child with a genetic condition and then to learn they could have known about it in advance.

Since the organization of obstetrics and gynecology in the Netherlands is structured in an integrated system similar to that of genetics, there are a limited number of sites where people can undergo prenatal testing. In the region of the country where I was located, the only place to have an amniocentesis was at the teaching hospital with which the clinical genetics center is affiliated. There are close ties between this ob/gyn practice and the genetics center. The senior obstetricians at the teaching hospital meet twice a week with staff from the clinical genetics center: once at the genetics center to discuss prenatal diagnoses, and once when a geneticist from the center attends the hospital's interdisciplinary prenatal meeting. The first of these meetings focuses on prenatal test results from the genetics laboratories. In the second meeting, physicians analyze cases in which they have found fetal problems through ultrasound.

Obstetricians are supposed to advise all women in the Netherlands with a maternal age indication for prenatal testing that such testing is available. The physicians staffing the ob/gyn practice at the teaching hospital handle these types of cases. If they find other factors suggesting an indication for prenatal testing in a particular case, they will usually discuss it with a clinical geneticist and often refer the individual for genetic counseling. A resident from the clinical genetics center goes to the teaching hospital for the prenatal outpatient clinic each week, in order to be available to consult on any cases in which such questions arise.

Postnatal Cases

The second group of patients the clinic deals with can be broadly defined as postnatal cases. These cases involve patients of all ages who fall into three subgroups. First are cases that overlap with the prenatal category in that the purpose of dealing with a postnatal case is to provide genetic counseling for people who are planning a pregnancy. A typical example was the case of Margreet and Joerdt. Margreet's sister is mentally disabled and lives in an institution. Margreet and her husband came to the clinic to inquire about their chances of having a child with the same problems as her sister. Since no definitive diagnosis was ever made for Margreet's sister's condition, the clinician's first step was to investigate Margreet's sister in the hope of diagnosing the problem and thereby being able to say something about the recurrence risk for Margreet and Joerdt.

The second subgroup of postnatal cases involves those in which a diagnosis is sought for a recognized problem. Sometimes these are cases in which a geneticist is called in after the birth of a child in whom anomalies are immediately noted and the pediatrician in charge does not know what to diagnose. Other cases involve adults who seek a diagnosis for the purpose of knowing the future consequences for themselves of a condition they have. Cases such as these include those of Jip, a clerk at a medical library, who has several dysmorphic features, and Carla, a thirty-two-year-old single performance artist, who has neurofibromatosis type 1 in her family. Jip had read an article about a genetic disorder that seemed to describe his morphogenic features, so he called the clinic in order to investigate whether this condition might explain his phenotype. Carla had arranged an appointment at the clinic because she knows that neurofibromatosis, a dominantly inherited genetic disorder, runs in her family; she wanted to know whether or not she had the disorder and, if so, what the consequences might be both for her own future and for that of any children she might have someday.

The third method through which postnatal cases come under the purview of the clinic is through satellite clinics. The center where I was based conducts regular outpatient clinics in the pediatric division of nearly a dozen regional hospitals. In these clinics they saw children whom the local pediatricians have referred for consultation with the geneticists.

Profession Building in the Clinics

The multiple ways people come into or are brought under the purview of clinical genetics centers in the Netherlands helps illuminate processes of profession building. While the examples above demonstrate the numerous paths that people take to the genetics centers, they also highlight how connections to ob/gyn practice and the practice of holding satellite clinics function as means of instantiating clinical genetics practices into both the broader medical profession and the wider social world. Dutch geneticists are in regular and frequent contact with obstetricians and gynecologists at the teaching hospitals, continually updating them on developments in genetics. An explicit purpose of the satellite clinics is to educate physicians beyond the teaching hospitals about what is made possible by advances in genetics in order to get them thinking in these terms.

The importance of educating general practitioners came up repeatedly in my conversations and interviews with geneticists. Physicians who lacked knowledge about genetics were sometimes construed as being outside the boundaries of the modern medical profession. When I asked a senior clinician, Diana, about changes in her field she introduced the problem of physicians not thinking in terms of genetics: "The general practitioner is not always very modern. . . . We try to tell the general practitioners, give them some courses, so they know much more about which people should be researched. . . . I've worked already thirteen years in this field, and I see there are still some doctors that do not understand . . . all the things that can be researched [by geneticists]." Diana wants to produce modern general practitioners. This is especially important because physicians who lack sufficient knowledge of genetics will not make the medical referrals to genetics centers that Diana thinks are necessary and appropriate. Extending the domain of genetics through the education of general practitioners, therefore, serves two purposes. First, it brings general practitioners within the ambit of the authoritative discourse of modern genetics. Second, it thereby ensures an increased flow of patients to genetics centers. As I elaborate in chapter 3, the patients themselves become modern subjects as they encounter the influence of modern genetics. The increased flow of patients, in turn, supplies genetic researchers with the chromosomes and clinical data necessary for the production of the type of genetic knowledge that drew the patients to the clinics in the first place. Educating general practitioners, therefore, establishes a sort of self-reinforcing "resource loop" to expand the domain of modern genetics.

Speaking about the medical profession more generally, Hans, another senior clinician, emphasized the need to define and establish genetics beyond the clinics. Clinical genetics, he noted, was still a very young specialty and did not "have a face yet." As a result, it was "not yet an integrated part of the thinking of doctors." He explained this by telling me that

> geneticists have to deal with the whole family and that way of thinking is not yet integrated into the whole medical profession . . . in the sense that . . . other physicians call a clinical geneticist and say, "Oh, I have a genetic disorder here and so the family is at risk perhaps, or perhaps

> not, can you check it?" . . . The doctor has to realize that he has a responsibility not only for the patient, but if he says that the patient has a hereditary disease he has to say it but it's not always said . . . many doctors say "you have this" and forget to say that it is hereditary. . . . The medical profession, medical doctors, [and] students have to change their attitude.

Hans is asserting that the medical profession is incomplete and even irresponsible to the extent that it fails to adopt the discourse and knowledge of modern genetics. For him, building the profession of medicine as a whole depends upon gaining wider recognition for the significance of the work of geneticists.

In chapter 3 I discuss how Dutch clinical practice of genetics normalizes the body. In Diana's and Hans's comments we see genetic practitioners also engaged in an attempt to normalize the medical profession itself by subjecting it to the discipline of genetic discourse. Many geneticists in the Netherlands share Walter Gilbert's belief that genetic knowledge is basic and primary to other forms of knowledge about the body. Locally, I found this sentiment clearly echoed in the repeated comment of one clinician that "all illness has a genetic component."

To the extent that geneticists establish their discipline as the key to understanding the human body, those who master the knowledge and vocabulary of genetics gain primary authority over all others who speak to questions involving the body. In other words, many of the geneticists I met in the Netherlands attempted to establish genes as a primary referent for all discussions about the body, health, and normality. As they succeed, those who speak to these questions gain authority and recognition only to the extent that they have mastered this knowledge and vocabulary.

Diagnosing and Researching Genetic Anomalies: The Clinic

Diagnosing genetic anomalies is a complex process involving numerous clinical and laboratory activities. Clinical analysis is an essential part of genetic practices for two reasons. First, for many genetic disorders no definitive laboratory tests are available to prove or disprove the presence of a particular condition. In these cases the only diagnosis available is a clinical diagnosis—one based on the expertise of physicians specializing in genetic disorders and involving physical exams, discussions with patients or their parents or both, and various laboratory tests aimed at suggesting, supporting, or ruling out potential diagnoses. Second, even when laboratory tests for a specific genetic disorder are available, they are conducted only when there is first a clinical indication for them. Much of the work involved in producing clinical diagnoses of genetic disorders occurs in a series of regular clinical meetings—primarily the audit, the interdisciplinary work group, and the National Consultation of Genetic Counseling (Landelijk Overleg Genetic Counseling).

The audit, a weekly meeting at the genetics center, took place one morning each week at 8:45 and was attended by all of the center's clinical geneticists, its physician specializing in cytogenetic diagnostics, and, for one year, their local ethnographer. The audit is made up of five main components. First, the meeting begins with each physician presenting a brief inventory of the new cases he or she has been assigned since the last meeting. Second, each physician provides a brief synopsis of each case for which he or she has concluded final genetic counseling during the previous week.

The third component comes in each audit when the clinic director asks if any of the physicians have questions. Although there were meetings at which nobody asked a question, at a typical meeting at least one person had a question to discuss with their colleagues. The questions often related to cases that never or rarely before had been encountered at the clinic, and they ranged from simple logistical issues about how to proceed with a case to deeply complex ethical questions or problems a physician had recently encountered and was trying to resolve.

During my fieldwork I found that several of the questions raised in the audit stemmed directly from recent scientific and medical developments, thereby throwing into relief the intersections among new medical and scientific technologies, genetics, the body, identity, kinship, and social values. In the chapters that follow I use several of the questions raised in the audit as a means of exploring these intersections and understanding the social conflicts and problems that scientists, physicians, and laypeople confronted in the context of new possibilities stemming from recent scientific and medical developments.

The presentation of patients who are difficult to diagnose makes up the fourth regular component of the audit. Unlike the discussion of questions, which occasionally involved a fairly detailed presentation aimed at resolving a logistical or ethical problem, the presentation of cases is intended to produce diagnoses. At a typical audit, between three and five cases were presented by physicians. One junior physician, Janneke, explained the purpose of the audit to me, saying that she believes its primary role has shifted over time. She said that initially the purpose of the audit was to enable the junior physicians and residents to draw on the expertise of the senior physicians in developing diagnoses. Geert, another junior physician, explained that the reason the clinicians "spend so much time consulting with each other has to do with the rarity of [genetic] disorders. . . . You can never get a lot of experience with things that have been seen only ten or twenty times in the world." Janneke's point also had to do with the rarity of some genetic disorders. When physicians at the clinic were unable to produce a diagnosis based on their own research and clinical experience or from consultation with one or two other colleagues, they would bring the case to the audit in the expectation that a physician with more experience in the field would be more likely to have seen, read about, or heard about a similar case and, therefore, be able to provide clues that might lead to a diagnosis. Janneke believes that new computer technologies have changed the relationship between junior and senior physicians. She explained that now that the clinic has access to a number of computerized medical databases, less experienced physicians usually are able to develop comprehensive lists of diagnostic possibilities on their own. Now, Janneke told me, the senior physicians "can rarely suggest something that we have not already considered." Thus, in her mind, the purpose of presenting difficult cases at the audit has taken on more of an educative function for clinicians at all levels of expertise. The physicians present the cases they are having difficulty diagnosing in hopes of obtaining help in producing a diagnosis, but they are also helping to educate each other by exposing each other to complicated cases.

The fifth, somewhat infrequent, component of the audit is made up of demonstrations. A demonstration, which also serves an educational function, is the presentation of a relatively unusual or complicated case that has been resolved. The purpose of demonstrations is to expose the clinic staff to the process of developing a diagnosis and, therefore, resolving such cases so they are aware of the types of clues that might point to such a diagnosis in cases they may see in the future.

The audit is run by the most senior clinician present, usually the clinic director. The highly ritualized nature of the meetings tends to obviate the need for the director explicitly to assert his authority in conducting the meetings. This does not mean he did not control the meetings. In fact, the director's authority suffused and guided the meetings. As friends are understood to know the nature of each other's identities, so too the clinicians were understood to know the nature of each other's authority. This produced an atmosphere conducive to Dutch egalitarian sensibilities even as hierarchy continues to operate. The people in the meetings are, in this regard, the same but different, a situation that resonates with the value I outlined in chapter 1 of simultaneously recognizing similarity and difference.

The explicit purpose of case presentations in the audit, both those presented for diagnosis and those given as demonstrations, is to take something that is perceived as a complicated medical problem involving numerous symptoms and fit it into a known category. As I explore these phenomena in the following chapters, I will demonstrate how the process of producing these categories and placing people within them reveals the explicit, complex, and contested dynamic of normalization at work in the production of clinical diagnoses of genetic anomalies. I will also illustrate the wide array of social, scientific, and medical knowledges that clinicians draw on in the diagnostic process. In the following chapters, I use my analysis of a series of case presentations from the audit as a means of elaborating the dynamic relationships among social life and medical and scientific practices as well as to illuminate the Dutchness of the practices I observed.

In addition to the weekly audit, another regular meeting was that of the Interdisciplinary Work Group on two Friday mornings every month. In this meeting the clinicians at the center were joined by neurological and metabolic specialists from the academic children's hospital with which the clinic is affiliated. The purpose of the meeting is to draw on the specialists' expertise in developing diagnoses, and the presentation of the cases is quite similar to the presentation of patients in the audit.

A third regular meeting that I draw on in my analysis is the monthly meeting of the National Consultation of Genetic Counseling (Landelijk Overleg Genetic Counseling), or LOG, as the staff of the genetics centers refer to it. This meeting, which took place on the third Thursday of every month, included virtually all of the physicians from all eight of the genetics centers in the country; it closely reproduces the form of the audit. The meeting begins with some administrative work, such as reviewing the minutes from the previous meeting. The second part of the meeting involves presentations, usually from geneticists working in research or diagnosis, meant to familiarize clinicians with recent developments in these areas. For example, when one of the centers was beginning a research project on X-linked forms of mental retardation, a geneticist from that center presented the goals of the study as well as the guidelines for including patients in the project. The goal of the presentation was to inform the clinicians about the project as well as to encourage them to refer appropriate patients for the research. The third component of the LOG is the presentation of complicated cases, cases that have resisted diagnosis in the various centers. These presentations are made in the same way that cases are presented at the audit. Occasionally the LOG questions had to do with logistics, but often the questions involved a physician asking if anyone present had, or had ever had, a patient with a particular disorder. In several instances physicians asked questions because they had patients with a disorder who wished to contact others with the same problem. If the physician raising a question at the LOG received a positive response from a colleague, the two would talk after the meeting.

One or more of these meetings was likely to play a role in developing diagnosis of the more complex or unusual cases physicians encountered at the clinic. Hans, a senior clinician, described the ideal path for producing diagnoses, saying that when an individual doctor is puzzled by a case she or he discusses it with a colleague. If they do not reach a satisfactory conclusion, they bring the case to the audit. If no solution is found at the audit, the clinic director will suggest that the case be taken to the LOG or to the interdisciplinary work group meeting and then, if no solution is found there, to the LOG.

In addition to the audit, the Interdisciplinary Work Group, and the LOG meetings, there are several other regular meetings that various members of the genetic centers' staff attend. Although I seldom draw on the content of these other meetings in my analysis, they are significant for understanding both the overall integration of Dutch genetics practice and its place within the wider medical establishment in the Netherlands. These include the two weekly meetings with the staff from the ob/gyn department at the teaching hospital and a weekly staff meeting involving clinicians as well as the social worker heading the psychosocial division and the senior staff of the diagnostic laboratories.

All of these meetings are important sites for understanding the social relations and social practices at work in the day-to-day operation of the genetics center and the development of clinical diagnoses. At the meetings, new developments in the field of genetics are discussed and incorporated into local practice. The meetings are also the sites where clinicians raise and attempt to resolve ethical questions they encounter in their work. The developments, questions, and resolutions that are elaborated in the context of the meetings open a window on the cultural production, reproduction, and change involved in recent and rapidly developing forms of medical and scientific knowledge in the Netherlands. In addition, these meetings serve as sites for the training and professionalization of physicians doing residencies in medical genetics.

DIAGNOSTIC LABORATORIES

The work of producing laboratory diagnoses of genetic disorders occurs in two types of diagnostic laboratories: a cytogenetic laboratory and a DNA laboratory. Chromosome analysis is conducted by laboratory technicians and cytogeneticists working in cytogenetic laboratories. These laboratories employ a wide range of technologies in discerning chromosome anomalies in both pre- and postnatal cases.

Before the laboratories can culture and analyze chromosomes, they must acquire the cells in which chromosomes are located. In the Netherlands, physicians were the primary people doing the work of obtaining these cells for the cytogenetic laboratories in the genetics centers. In postnatal cases these cells were drawn most frequently from blood or from small skin biopsies. Acquiring cells for chromosome analysis in prenatal cases, by contrast, requires access to fetal cells, access gained through either amniocentesis or chorion villus sampling. Amniocentesis, introduced in 1960, involves inserting a needle through a pregnant woman's abdominal wall, uterus, and the membranes surrounding the fetus. The physician carrying out the amniocentesis then pumps out a sample of amniotic fluid through the needle. This amniotic fluid contains fetal cells that have been sloughed off during fetal development and can be cultured for chromosome analysis. Chorion villus sampling is basically a biopsy of the placenta. This sampling can be done either through a pregnant woman's abdominal wall, as in amniocentesis, or through the insertion of a needle through a catheter, or tube, inserted through the cervix. Since the cells of the placenta also come from the original zygote, itself formed from the combination of egg and sperm, physicians expect that chromosome anomalies in the cells of a fetus would also be found in those of the placenta.

The primary differences between amniocentesis and chorion villus sampling have to do with timing and accuracy. Amniocentesis was usually done around the sixteenth week of pregnancy. The culturing and analysis of the cells drawn during the procedure required about two weeks, though this could sometimes stretch out to as many as four weeks depending on how fast the cells in a particular case grew. Thus, amniocentesis test results come back well into the second trimester of a pregnancy. Chorion villus sampling can be conducted between the ninth and twelfth week of a pregnancy, meaning that the test results can come back within the first trimester of a pregnancy. The results, however, may not be as accurate as those afforded by amniocentesis. In some cases, a physician might have reason to suspect that a chromosome anomaly found through chorion villus sampling may be confined to the placenta. In these cases an individual may choose to wait and undergo amniocentesis for more accurate test results.

Once a physician acquires the necessary cells, she sends them to a cytogenetic laboratory for culturing and analysis. At the lab, technicians and cytogeneticists use a variety of chemicals, stains, temperature changes, photography, and computers to develop the test results. The technicians and cytogeneticists represent the results of culturing and analysis in the visual images of chromosomes known as karyotypes. They communicate the results back to patients via physicians in the clinics.

In addition to the standard chromosome analysis that results in karyotypes, some laboratories used a process called fluorescent in situ hybridization (FISH) when looking for certain specific chromosome anomalies. This process involves special technologies in the form of expertise, fluorescent stains, probes that attach to distinct sites on specific chromosomes, and photo and computer equipment able to capture the fluorescent pictures produced by FISH. FISH was a relatively new technology used to identify certain chromosome anomalies so subtle that they cannot be observed in a standard karyotype. The resulting pictures are quite dramatic, showing fluorescent chromosomes with a few dots of a second fluorescent color all glowing against a black background. At the time of my fieldwork only two laboratories in the Netherlands had FISH capabilities. As in the case of standard karyotypes, the information developed through FISH is communicated back to patients via clinic physicians.

In DNA laboratories, laboratory technicians and molecular biologists analyze genetic anomalies at the level of DNA. The cells needed for analysis are obtained in the same way as the cells for chromosome analysis—in most postnatal cases from blood or skin biopsy and in prenatal cases from amniocentesis or chorion villus sampling. Rather than looking at the gross structure and number of whole chromosomes, as cytogeneticists in cytogenetic labs do, molecular biologists in DNA laboratories analyze chromosomes for far smaller variations at the level of DNA. These include the presence or absence of one or more base pairs in a specific gene, such as in cystic fibrosis, or multiple repeats of certain base pairs at a particular point within a gene, such as in cases like fragile-X syndrome, Huntington's disease, and myotonic dystrophy, which are linked to a series of repeats of specific base pairs within a gene.

The technologies available for laboratory diagnosis can also be used as a means of educating general practitioners and of drawing patients to the clinics. On several occasions one senior clinician instructed the resident accompanying him to a satellite clinic in a regional hospital to include FISH pictures with the materials they normally take to satellite clinics—patient folders, camera and film for photographing patients, and test tubes for collecting blood to bring back to the diagnostic laboratories. Between seeing patients or at the end of a satellite clinic, the geneticists confer with the local physicians and briefly summarize what they have found as well as arrange a date for their next visit to the hospital. At the two satellite clinics I attended to which geneticists brought FISH photographs, the senior clinician, Hans, used the discussion time to introduce the photos to the local physicians. Because the local physicians had not previously seen such images, Hans explained that FISH stands for fluorescent in situ hybridization. While showing the photographs, he discussed the use of a fluorescent probe, explaining that this diagnostic process was very fast and that only two places in the whole of the Netherlands have the capability to do it. He went on to outline what types of syndromes they could use the technology to diagnose as well as what symptoms might be indications for such tests. He concluded by telling the physicians that geneticists are always discovering more things they can test for through this technology.

RESEARCH LABORATORIES

Each of the genetics centers has an active program in human genetics research, though these vary in size from center to center. These laboratories are staffed by laboratory technicians, doctoral students, and molecular biologists conducting research and publishing their results in major international scientific journals. The primary relationship between the research programs and the clinical practices that is significant for my analysis is that the clinical practices provide a basic material component to the research laboratories—the DNA of individuals with specific genetic anomalies. When a laboratory begins research in a new area, researchers from the project will arrange to make a presentation at the LOG as a means of informing clinicians about the project. One component of such presentations is a description of the guidelines for determining the kinds of patients who might be candidates for inclusion in the study. The researchers circulate the guidelines in writing to the centers so that the clinicians have a copy of them available in their offices. The organization of the centers in the Netherlands, where eight clinical centers deal with all suspected genetic conditions in the country that are referred for investigation, makes for an efficient—and the primary—source of DNA from appropriate candidates within the country.[8]

The organization of the eight Dutch genetics centers highlights the deep integration of the institutional structures in which genetic research and clinical practices take place in the Netherlands. This integration illuminates the many social relationships through which knowledge about genetics is produced. The genetics centers are primary sites for encounters between medical professionals, researchers, and the general population as they mutually negotiate the meaning of genetic knowledge in the Netherlands. The centralized organization and production of medical and scientific knowledge about genetics in the Netherlands have two important consequences for these encounters. First, they build a coherent and unified profession of geneticists throughout the country. Second, they thereby enhance the production of an authoritative discourse about the nature and meaning of genetic knowledge. The potent institutional framework through which Dutch people encounter authoritative knowledge about genetics highlights the particularly Dutch organization of a major medical and scientific endeavor. It also illuminates the myriad social relationships involved in producing and managing scientific knowledge about the body in the Dutch practice of medical genetics. It is in this institutional framework that knowledge about genetics is produced, interpreted, translated, and consumed in the Dutch context.

Chapter Three

THE SOCIAL AND CLINICAL PRODUCTION OF ORDINARINESS

This chapter explores how genetic knowledge is deployed in daily life in the Netherlands through the lens of the weekly meeting of the genetic center's physicians known as the audit. This clinical practice is produced through a convergence of medical practitioners' desire to identify and pathologize difference, the Dutch practice of recognizing and bounding difference, and Dutch values about ordinariness. The audit combines advances in genetics with normalizing processes to enable both the production and the bounding of differences associated with genetic conditions.

Focusing on the audit serves three purposes. First, it illustrates the dynamic nature of the multiple activities creating the world of clinical or medical genetics in the Netherlands. Second, it highlights how the production of clinical genetic diagnoses normalizes both the interior and the exterior of the body in this clinical setting. Third, it demonstrates how distinctively Dutch social ideals of tolerance, ordinariness, and social life intertwine with the production of scientific and medical knowledge through daily practice. Taken together, these phenomena show the extent to which the structures of normality are taken for granted in everyday practice including biomedical practice.

The unusually explicit normalizing processes through which Dutch medical geneticists diagnose and categorize people with genetic conditions powerfully resonate with broader social imperatives associated with pillarization as a worldview. I argue that a central component of Dutch genetics practice involves bounding and containing difference through a process of fitting people into categories in which they can be perceived as ordinary and, hence, acceptable members of Dutch society. My intent is not to question how diagnoses are produced; my argument should not be taken as suggesting that Dutch genetics is less rigorous than medical or scientific practices elsewhere. Rather, the analysis demonstrates that even as scientific and medical practices are connected with a global cosmopolitan community of professionals, they also remain imbricated in locally specific cultural commitments. As products of lived experience they can never be wholly divorced from their local contexts. The local specificity of genetics in the Netherlands points to the broad need to develop understandings of modernity, science, and the West that are more nuanced than those presently recognized.

In the practices of normalization and genetics we will see elaborated a biomedical counterpoint to the social construct of pillarization. In this context, pillarization offers a culturally salient framework for talking about difference that does not take geneticists down the path to the hereditarian discourses that fueled the racism of the Nazis, of which they are all too aware.

NORMALIZATION AND THE PRODUCTION OF TRUTH

Until I began research on genetics in the Netherlands I had always thought of normalization as a subtle process. I had been influenced by Foucault's depiction of the techniques of normalization as "capillary" (Foucault 1980a:84; 1980b:96) and involving "small acts . . . with a great power of diffusion, subtle arrangements, apparently innocent . . . mechanisms" (Foucault 1979:139). Foucault (1980b) opposes these delicate terms to the explicitness of repressive forms of power based on a binary opposition between permitted and forbidden behavior. This oppostion in his descriptions of modern and classical forms of power suggests that he himself conceived of modern power as a subtle process. In Dutch genetic centers normalization is far from subtle. It is, rather, explicit and open, even obvious. Despite its openness, normalization in the Dutch context is still complex and contested. The Dutch case illustrates that normalization can be a pluralistic process in which multiple categories of "normal" exist simultaneously, so that normalization takes on a multivalent quality. In the context of Dutch genetic practices, the category of normal exists at two levels. First, there is a primary, unmarked category of genetically normal that comes into play in relation to abnormality. The issue of a person's genetic normality is articulated only when presented in contrast to someone with a genetic disorder. Second, rather than simply classifying an individual with a genetic anomaly as abnormal, clinical geneticists go to great lengths to fit the person into a scientifically or medically defined category in which she or he may be perceived as normal. This process produces multiple categories of normal, each of which is contained within distinctly classified genetic conditions. Paradoxically, then, Dutch genetic practice classifies individuals with genetic disorders as both abnormal in relation to an unmarked category of normal and normal within their own category of genetic difference. Like the process of pillarization, the process of normalization in the Dutch clinics bounds and contains difference.

Medical geneticists, research geneticists, and laypeople, both inside and outside the clinical setting, mutually negotiate categories of normal, producing new categories within which people with genetic syndromes or anomalies fit. In order to accomplish this task a range of actors collect, create, and interpret evidence that speaks to a community of experts to produce truth about the body (Latour 1987). These explicit, formidable processes do not simply produce medical diagnoses, but also enable the Dutch social ideal of ordinariness.

THE AUDIT

The audit took place one morning each week in the library of the clinic. The center's four clinical geneticists, its eight physicians doing residencies in clinical genetics, and its physician specializing in cytogenetic diagnostics all attended the meeting.[1] The beginning of the meeting was punctuated by the noise of pouring coffee and stirring spoons. The director of the clinic would ask one of the physicians to begin, and each physician would, in turn, give a brief summary of the new cases assigned to them during the previous week. Each physician would then give a summary of the patients he or she had met at a final genetic counseling appointment during the previous week. In this section I focus on the concluding component of the audit: the presentation of patients the clinicians found difficult to diagnose. At a typical audit two to five patients would be presented in this segment of the meeting.

The presentations adhered to a standard ritualized format. They began with the clinician in charge of the case placing a transparency with a case summary on the overhead projector (see text below).[2]

2nd child of healthy parents, 2 healthy sibs,
mother is pregnant again

pregnancy and birth nothing unusual, G 2540g

family: Mother + Brother surgery for crossed eyes
Father cleft lip (palate?)

dysmorphic features: asymm. face
pre auricular tag R

short neck

pediatrician's diagnosis: Goldendhar syndrome

development: slow growth (p3)

delayed speech

surgery for crossed eyes

surgery for hip displacement (motor skills also a little slow)

research: L101.6cm (<p3), span 96 cm, OFC 47.5cm (p3)

asymm. face at disadvantage of right

short neck causing constraint

uneven shoulder R > L

restricted motion R arm (also a little thinner)

left side more strongly developed

supplemental: ultrasound kidneys nothing unusual

chromosome research 46, XX

ultrasound skull nothing unusual

X-SC nothing unusual

X-CSC very extensive deviation with wide () distance,
missing vertebrae parts and a congenital half vertebrae?

X-hips: displacement on both sides

X-arms/legs nothing unusual

still: hearing test?

X-cwk parents?

Klippel-Feil?

The clinician then provided an oral summary of the information in the overhead transparency; finally, he would present slides of the individual for whom he was seeking a diagnosis. Bart, the junior clinician in charge of one case, summarized it thus:

> This is the second child of healthy parents, well the father has a cleft lip and perhaps a cleft palate but he doesn't want to talk about it. There have been two miscarriages. There are two healthy sibs, and the mother is again pregnant so they aren't really interested in recurrence risk. Pregnancy and birth were normal. She has had surgery for crossed eyes but so have her mother and one brother. The father has a cleft lip, and perhaps a cleft palate, but he doesn't want talk about it and doesn't fill it in on the family form. She has an asymmetrical face, a pre-auricular tag on the right side, and a short neck.[3] The pediatrician suggests Goldenhar, but I think that maybe it is Klippel-Feil. Her height is in the p3, and her speech is delayed.[4] She has had surgery for the crossed eyes and also for hip displacement, and her motor skills are also a bit slow. You can see for yourself what the research shows . . . the only thing that the X-rays showed was the vertebrae abnormalities. The only other things that I could think might show something would be a hearing test or an X-ray [of the cervical area] of the parents' [spinal columns (X-CWK)].

Toward the end of the summary the presenting clinician would ask one of his colleagues to turn on the slide projector and show slides of the individual or individuals in question. The slides were normally (though not always) photographs of the individual taken by the physicians themselves in a clinical setting. Occasionally images were provided by the family; these were generally photos taken at family gatherings and holidays.

Following the case presentation, the physicians would ask questions and offer comments on the pictures. The questions often related to individual and family medical histories and are intended to suggest possible diagnoses. The pictures of Ineke, whose case I have presented above, showed a little five-year-old girl with shoulder-length blond hair, blue eyes, and a sweet smile. They also showed that she has an extremely short neck and one shoulder noticeably higher than the other, the combination of which looks as if it inhibits her neck and arm mobility. In presenting Ineke's case, Bart showed photos of her front, both sides, and her hands; he also showed one of her back from her shoulders to just below her knees in which her underwear is pulled down below her hips.

Upon viewing the pictures, the physicians would begin to discuss the features of the person in the photographs. They inquired further about measurements and test results and ultimately proposed possible diagnoses. In discussing patients' photos, the physicians returned repeatedly to specific themes, including the patient's hairline and color, the shape of the nose, the placement of the ears, and whether the face was coarse. There were almost always questions and comments about whether patients resemble other members of their family—if they looked as though they belonged to them—in terms of both physical features and intelligence.

The photos and physical examinations of patients emphasize the outside of the body, focusing on both specific morphogenic features and overall appearance as means of categorizing difference associated with genetic pathology in a particular family. At the same time, reports of the inside of the body and its functioning included in the presentations serve to turn the physician's gaze to the interior of the body. These reports include results from technologies such as X-rays, ultrasound, metabolic screening, CAT-scans, and chromosome and DNA analysis. Like the other presentation materials, these test results were most often represented in the audit as visual images. As the presentation of Ineke and her family demonstrates, the clinicians expect these internal features of the body will also show family resemblances. The audit, therefore, provides the opportunity to normalize both the interior and exterior of the body.[5]

The explicit purpose of the case presentations is to take something perceived as a complicated medical problem involving a variety of symptoms —including morphogenic features—and fit it into a known category. This process takes people who do not fit a particular category of normal—because they are severely dysmorphic or mentally disabled or marked by milder symptoms that somehow set them apart—and place them into a separate category in which they may be considered typical for that category. Thus, even if one does not look like the members of one's family, clinicians expect to find a category into which a person with a genetic condition fits. This is not to say that such categories deny anomalies. Rather, the categories bound and contain difference by determining a category in which the characteristics associated with that condition are ordinary.

Ineke, like many of the people whose photographic images I saw in the audit, may not have looked like her family members or otherwise appear normal, and her body may have functioned in ways that set her outside the norm, but in the category of Klippel-Feil syndrome she might be securely settled in the range of what was considered typical for that category. Consider also the following cases—each was presented in the audit, and each attests to the emphasis that the clinicians at the genetics centers place on similarities within families:

Case 1: Discussing possible diagnoses for a child in an audit, the clinicians were having trouble developing one that made sense to them. One clinician suggested the possibility of Stickler syndrome, but Hans, the physician in charge of the case, said he had already ruled out Stickler. He said, "She's significantly retarded compared to the parents and sibs." One of his colleagues responded, "So it is probably something." To which Hans replied, "Yes, but what?"

Case 2: Presenting a case in which she diagnosed Angelman syndrome, Lauren, a junior clinician, described some of the symptoms she found in the child—"serious retardation, microcephaly, epilepsy . . . she is blonder than her sibs."

Case 3: Presenting a case for diagnosis, Bart explained that he had a case in which the parents had come for genetic counseling before the wife became pregnant again because one of their two children exhibits a combination of medical problems. In order to be able to counsel the couple as accurately as possible Bart is trying to diagnose the child in the context of the family's medical history. One grandfather and that grandfather's brother have a muscular disease. Records indicate that "pregnancy and delivery were normal." Bart described the symptoms: "From one year on there was epilepsy that was very difficult to treat and even with treatment it became worse—sometimes twenty episodes a day . . . he has trouble walking . . . maybe there is muscular disease in addition to epilepsy? . . . the MRI of his muscles is inconclusive . . . chromosomes are normal and negative for fragile-X . . . he is short and fat but so are the parents . . . parents are also not very bright—that is even an understatement . . . but even given the level of these parents, this child drops out because he has a brother who is one year younger who can do everything better."

Case 4: Presenting a case of a child for diagnosis, a clinician describes some history and symptoms—"Premature birth, very small . . . she has a long philtrum as the mother does." In the discussion following the presentation one clinician asked how "retarded" the child is. The presenting clinician responded, "Not very." When another clinician commented on the small circumference of the child's head, the presenting clinician said, "Mother said that a lot of people in the family have small heads." Then, following up on the issue of mental retardation the clinician said, "A lot of family members have learning problems." In looking at the photos accompanying the case, the clinicians focused in particular on a picture of the girl at the age of three years and four months with her brother, one of them commenting, "He [the brother] is completely different."

Each of these cases highlights the emphasis clinicians place on family similarity as a sign of normalcy and, conversely, their understanding that morphogenic dissimilarity within a family may be a sign of pathology. The emphasis on family similarity and difference underscores how, under the rubric of genetics, the family itself becomes a normalizing category.

The physicians' discussion following the presentation of Ineke's case centered both on fitting her external morphogenic features and her internal features and bodily processes into a coherent picture that would place her within a known, bounded category. At the end of his presentation of Ineke's case, Bart explained that while the pediatrician at the regional hospital proposed a diagnosis of Goldenhar syndrome, he did not think Ineke had Goldenhar, but Klippel-Feil syndrome, a syndrome, like Goldenhar, in the spectrum of cervico-oculo-acoustic disorders. In concluding his presentation, Bart told his colleagues that the two disorders were "part of the same spectrum, so the distinction is a little bit semantic" but that from his point of view Ineke "has enough [symptoms] for Klippel-Feil and not enough for Goldenhar."

There was a moment of silence while the clinicians looked at the slides. Although, as I discuss below, the physicians regard these photographs as an essential component of the diagnostic process they were considered not very useful in Ineke's case. For example, a senior clinician began the discussion by saying, "Well, it is really a radiological diagnosis," meaning that the precise genetic basis of the disorder is unknown, so evidence for a Klippel-Feil diagnosis comes from X-rays. Another physician commented, "If you say that Klippel-Feil is one in a spectrum of O.A.W., then this is Klippel-Feil.[6] If you don't, then you say that it is all O.A.W. and Klippel-Feil doesn't exist." The presenting physician asked, "What do you suggest for her prognosis?" and was told, "Advise them to come back when she is twenty. Then there will be more children of people with O.A.W. [described in the literature]." Later, Bart explained to me that Goldenhar and Klippel-Feil were part of the same spectrum and that he did not think the distinction was very important to the parents.

The contrast between the physicians' concerns and Bart's perception of the parents' concerns is striking. Bart explained that the parents were interested not in recurrence risk, but in the specific consequences for Ineke. The physicians, however, were interested not only in Ineke, but also in more abstract questions having to do with the construction and understanding of diagnostic categories. The suggestion that Klippel-Feil might not exist indicates that geneticists continue to consider certain categories of genetic diagnoses as open and unfixed. Another physician spoke to this issue when she told me, "Probably [what we now call a] syndrome is not a syndrome. A lot of children who look alike are put together and maybe in ten years' time, or twenty, or one hundred, it does not matter, we can distinguish between them because look-alikes might be two syndromes." These comments reveal the open and evolving nature of genetic categories. At the same time, Ineke's case demonstrates the clinical emphasis on fitting people into categories. There is a tension between the open, evolving nature of genetic knowledge and the urge to classify, bound, and contain difference. The tension, however, is only apparent. Geneticists are willing to live with the uncertainty of evolving categories. They do not, however, question the ultimate imperative to categorize. They may question categories but not categorization.

SYNDROMES AND THE PROBLEM OF "ENOUGH"

One important and powerful means of categorization involves the concept of a syndrome. Many of the disorders the clinicians deal with in the audit are known as syndromes. Syndromes are the basic categories by which geneticists assemble individuals with distinct symptoms into larger groups. Both the concept and the content of a syndrome are complicated. Geneticists at the center repeatedly described the concept of a syndrome to me by explaining that they define a syndrome as "a combination of symptoms that come together in combination more often than could be expected by chance."

The clinical imperative to find a syndrome is illustrated by a case presentation in the audit involving a child with many symptoms that did not fit a single diagnosis. In this case, many, but not all, of the symptoms could be explained by an infection the mother had or may have had during pregnancy. The geneticists and residents began discussing the case after it was presented in the audit. At one point two clinicians suggested the possibility of there being multiple causes for the multiple symptoms. The clinic director immediately interrupted them to state what he considers a central tenet of genetic diagnosis: "One child, one problem, most of the time." The director here urged the other clinicians to take a complicated combination of symptoms and fit them together into a single, known category. His comments also indicate the desire to limit individuals to one syndrome, thereby bounding and containing the symptoms that mark an individual as different.

The content of a syndrome is complicated by geneticists' understanding that although all people with a specific syndrome share certain symptoms, not every syndrome-bearer necessarily possesses all the symptoms associated with that syndrome. This understanding of what constitutes a syndrome was made clear in the audit when the presenting physician included a transparency that listed the major symptoms associated with a particular syndrome along with the percentage of syndrome-bearers in whom each of those symptoms was manifest. The presenting physician then put a plus or minus sign next to each symptom listed in the table to indicate whether the patient they were presenting had that symptom (fig. 1). This context offers insight into the questions the clinicians raise about how many symptoms are "enough" to constitute a syndrome. Just as Bart argued that Ineke had enough for Klippel-Feil but not for Goldenhar, at the end of another case presentation a physician asked her colleagues, "Is it enough for Stickler syndrome?" Although Bart did not use a table of symptoms in his presentation of Ineke's case, he did have one in her file, having used it to help determine whether Ineke had enough symptoms for Goldenhar or Klippel-Feil. The concept of enough here involves a sort of diagnostic arithmetic in which symptoms are tallied on a genetic scorecard.

The listings of these tables may communicate far more than just the symptoms indicative of a particular disorder. For example, ethnicity and nationality may also be introduced into them. In a presentation of two brothers in whom the physician suspected Cohen syndrome, a transparency of a table listing the "Major Clinical Signs of the Cohen Syndrome" was used. The article from which the transparency was copied (Kondo et al. 1990) compared the manifestation of symptoms by ethnicity and nationality, using a group of Jewish people with Cohen, a group of Finnish

TABLE I. Major Clinical Signs of the Cohen Syndrome

Clinical signs	Jewish (n=39) (%)	Finnish (n=6) (%)	Others[a] (n=42) (%)	Case 1	Case 2
Growth and development					
Mental deficiency	39(100)	6(100)	41(98)	+	+
Short stature	5(13)	6(100)	30(71)	+	±
Tall stature	8(21)	0(0)		–	–
Mild truncal obesity	16(41)	6(100)	29(69)	–	–
Mild hypotonia	33(85)	6(100)	35(83)	+	+
Delayed puberty	13(33)	3/4(75)		+	?
Cheerful disposition	39(100)	6(100)		+	+
Craniofacial					
Microcephaly		6(100)	22(52)	+	+
Down slanted eyes	34(87)	1(17)	18(43)	–	–
High nasal bridge	39(100)	6(100)	34(81)	–	–
Short philtrum	39(100)	6(100)	32(76)	+	+
Prominent upper central incisors	38(97)	3(50)	29(69)	+	+
Open mouth	37(95)	6(100)	33(79)	+	+
High/narrow palate	39(100)	6(100)	31(74)	+	+
Maxillary hypoplasia	34(87)	5(83)	23(55)	+	+
Micrognathia	27(69)	5(83)	33(79)	+	+
Malformed ears	25(64)	6(100)		+	+
Limbs					
Long/narrow hands and feet	39(100)	6(100)	37(88)	+	+
Hyperextensible joints	39(100)	6(100)	23(55)	+	+
Mild syndactyly	29(74)	6(100)	10(24)	+	+
Ophthalmologic					
Myopia	11(28)	6(100)	16(38)	+	+
Strabismus		3(50)	22(52)	–	–
Mottled retina	1(3)	6(100)	15(36)	+	+

[a]Cohen et al. [1973], Carey and Hall [1978], Balestrazzi et al. [1980], Fryns and Van den Berghe [1981], Ferré et al. [1982], Goecke et al. [1982], de Toni et al. [1982], Doyard and Mattei [1984], North et al. [1985], Resnick et al. [1986], Young and Moore [1987], Rizzo et al. [1987], Zetler et al. [1987], Méhes et al. [1988], Nambu et al. [1988].

1. Major clinical signs of Cohen syndrome.

people with Cohen, and a group of others. According to this table, 100 percent of Jewish and Finnish people with Cohen syndrome have, among other symptoms, "mental deficiency," "cheerful disposition," "high nasal bridge," "short philtrum," and "long/narrow hands and feet," but only a fraction of the others diagnosed with Cohen share these symptoms. In the group of others, 98 percent have "mental deficiency," 34 percent a "high nasal bridge," and 32 percent a "short philtrum"; their disposition was not reported.[7] Furthermore, not all Jewish and Finnish bearers of Cohen are alike. Of the Jewish cohort, 13 percent exhibit "short stature" and 21 percent "tall stature" (presumably the other 66 percent fall within a normal height range), while 100 percent of the Finnish group display "short stature." In this case presentation the physician also used a second chart listing symptoms associated with Cohen syndrome, comparing the expression of symptoms, including "cheerful disposition," in sixty-one people whose ethnicities were unspecified. This example of the clinical symptoms associated with Cohen syndrome highlights the complexity of establishing what constitutes enough symptoms to allow positive diagnosis of a syndrome. Not all syndrome-bearers share all the symptoms associated with a particular syndrome, and such factors as race, ethnicity, and disposition can come into play in the production of genetic diagnoses.[8]

Once a clinician has established whether a patient has enough symptoms to allow a diagnosis, she or he may be able to offer patients and their families some predictions. Such predictions, however, are complicated by the understanding that syndrome-bearers may manifest different symptoms that also range in severity. Syndromes do not exist objectively as such in nature. They are named and created by geneticists to help them in the processes of diagnosis and categorization. The relational and constructed nature of syndromes is illustrated by a table that establishes diagnostic criteria for Noonan syndrome developed as part of a research protocol at one of the Dutch genetics centers. The table was distributed to clinicians in each of the genetics centers to make them aware of the criteria for involving a particular patient or family in a new research program on Noonan syndrome.

Noonan syndrome, primarily known as an X-linked form of mental retardation, is marked by, among other symptoms, particular facial, cardiac, and skeletal features. According to the table there are two types of Noonan syndrome, type A and type B. An individual is diagnosed with Noonan only if he has the proper combination of the symptoms associated with it. The potentially relational nature of constructing a syndrome is revealed by the table's designation of a symptom based on family history. Imagine the following scenario: Two brothers, Vincent and Theo, come to a clinic, where Vincent is diagnosed with Noonan syndrome. He has facial features typical for Noonan, and he is short, falling below the third percentile for height. Theo has a face that is suggestive of Noonan syndrome, and he falls below the tenth percentile for height. Because Vincent's symptoms are enough to diagnose him with definite Noonan syndrome, Theo need manifest only one more symptom from column B to also satisfy the criteria for a definite diagnosis of Noonan syndrome. Hitherto, Theo had always been considered a bit slow at school, but no one ever thought much of it. The question may then emerge as to whether his mental slowness is actually a form of mental retardation. If so, Theo will satisfy the final criterion necessary to establish a diagnosis of Noonan. In this context, the clinical dynamic of diagnosis may recharacterize Theo's mental slowness as retardation because of his relation to Vincent. In any event, the clinicians may use the relationship between the brothers to produce the diagnosis.

The concept and content of syndromes help geneticists to organize, bound, and contain the myriad symptoms encountered in daily clinical practice. Without a syndrome, there is no diagnosis, only a description of symptoms. The idea of one child, one syndrome, most of the time reflects

TABLE 1. Diagnostic Criteria for Noonan Syndrome

Feature	A = major	B = minor
1. Facial	Typical face	Suggestive face
2. Cardiac	Valv. PS and/or Typical ECG	Other defect
3. Height	$<3^{rd}$ centile	$<10^{th}$ centile
4. Chest wall	Pectus carinatus/Excavatum	Broad thorax
5. Family history	First degree Relative definite NS	First degree Relative sugg. NS
6. Other	All three (males): Mental retardation Cryptorchidism Lymphatic dysplasia	One of: Mental retardation Cryptorchidism Lymphatic dysplasia

Definite Noonan Syndrome: 1A plus * one of 2A–6A or * two of 2B–6B
1B plus * two of 2A–6A or * three of 2B–6B

the idea that for every person there is a proper category. Such an idea also implies that, ultimately, diverse symptoms should fit together and relate to each other in a coherent, manageable manner.

MAKING SENSE OF THE AUDIT

The traditional goal of clinical work is to transform medical complexities into solvable, treatable, and curable problems. This clinical goal is often complicated in the case of medical genetics, a field in which the availability of treatment varies widely depending on the type of disorder. The primary goal of presenting patients at the audit is to arrive at diagnoses. The enormous expense involved in terms of the number of people, the time, and the equipment that can be called upon in a complicated case indicates a deep commitment to the diagnosing of genetic disorders. Diagnosis might seem an obvious and routine step on the path to treatment and cure. Yet, despite all the publicity about advances in genetics, there are no cures for genetic syndromes, and in many cases, such as Ineke's, there are no available treatments.

Clinicians regularly articulated four intersecting factors in discussing the importance of accurately diagnosing genetic disorders. First, accurate diagnosis may well guide palliative treatment and provide significant information about prognosis. In cases such as the connective tissue disorder known as Marfan syndrome, the use of beta blockers, anticoagulants, and cardiac surgery offers significant positive benefits in terms of quality and extension of life (Heath 1998). In other cases few palliative treatments are available, but diagnosis might allow predictions about the future consequences of a particular disorder for a patient or her or his family. In still other cases clinicians may be able to make general relational predictions about how severe the symptoms will be in a particular individual. Such predictions are, as I elaborate below, relational because they depend on a clinician having knowledge about the structure of the gene and the severity of symptoms in the individual's affected relatives. The availability of this kind of relational information depends upon the willingness of family members to participate in clinical and laboratory analysis. In yet other cases predictions are impossible because the cause of the variation in symptoms is unknown.

Second, diagnosis may provide valued reproductive information about recurrence risk and the possibility of prenatal testing to people affected with a certain syndrome or to parents who have had a child with a genetic condition. Diagnosis often reveals the inheritance pattern of a syndrome, enabling clinicians to make predictions about recurrence risk for the syndrome-bearer or for the parents of a child with a genetic condition or both. In presenting one case for diagnosis, Bart explained to his colleagues that he was seeking diagnosis because "these people have a deep desire to have children [*ernstige kinderwens*], thus they want a diagnosis and recurrence risk."

The most common recurrence risk predictions are based on geneticists' understandings of simple Mendelian patterns of autosomal dominant, autosomal recessive, and sex-linked inheritance. Such predictions, however, are not always possible. Indeed, in Ineke's case, her parents had brought her to the clinic to investigate the consequences of her condition for her. From the clinicians' point of view those consequences necessarily involve Ineke's own possible future reproduction. It is for this reason they suggest that Ineke come back in fifteen years, by which time they hope to have learned more about the reproductive consequences for people diagnosed with disorders in the O.A.W. spectrum. Diagnosis also allows clinicians to inform people whether a prenatal genetic test is available for the syndrome for which they are at risk. Such tests can inform people about whether or not the fetus they carry is affected with a disorder. They are rarely informative about the severity of the symptoms should the parents continue the pregnancy and have a child.

The case of Peter and Marianne, the couple who were in their late thirties and used in vitro fertilization to achieve Marianne's pregnancy, underlines the ambiguity and stress that can be involved in prenatal diagnosis as well as the relational nature of some genetic diagnoses. I first met Peter and Marianne when they agreed to allow me to sit in on their genetic counseling sessions with Iris, one of the clinic's junior physicians. The sessions took place at the regular prenatal outpatient clinic at the local teaching hospital. As I mentioned in chapter 2, Marianne has myotonic dystrophy, and the couple wanted prenatal testing because they had decided to have a child only if it did not have the gene for myotonic dystrophy. Marianne had learned that she carried the gene for myotonic dystrophy when she had participated in testing with other members of her family several years before becoming pregnant. At the time the family was tested the gene for myotonic dystrophy had not been found and nothing was known about how it functioned. Researchers had determined, however, which chromosome the gene was on and therefore were able to conduct testing based on coupling research—a type of research through which they could determine, by looking at the chromosomes of a number of both affected and unaffected family members, which family members carried the allele with the defective gene. Marianne has very mild symptoms and did not appear at all dysmorphic or disabled when I met her, something that Iris commented on in discussing the case with me after the appointment.

When Peter and Marianne came into the office for their prenatal appointment they were clearly anxious. Marianne's eyes welled up with tears several times during the appointment. Iris explained to them that myotonic dystrophy is a dominant condition, so there was a 50 percent chance that their baby would have the gene. She told them that in the years since Marianne had done carrier testing based on coupling research, researchers had found the gene and learned a bit more about it. The problem, she explained, has to do with an expansion in the number of times a set of three base pairs repeat in the gene. People who are affected have a certain number of repeats, and the number of repeats in an individual's gene bears a relationship to how serious their symptoms are. She further explained that if the fetus carries the gene and has repeats that number in the same range as in Marianne's gene, they could expect that if they decided to carry the fetus to term, the baby would likely have symptoms similar to Marianne's. Iris said that testing might show that the fetus has many more repeats, and in that case they could expect the symptoms would be somewhat more serious than Marianne's. She also told them that in very rare cases a baby had fewer repeats than the parent.

The information seemed to overwhelm Marianne. With tears in her eyes, she said she found it more difficult to have this nuanced information than if she had known just whether the fetus had the gene. Iris pointed out to Marianne that she had quite mild symptoms and asked her if she knew what she would do if the results showed that the baby had the same number of repeats as she did. After a pause, Marianne said they would terminate the pregnancy because she wanted a healthy child (*gezond kind*). She added that she could not predict what would happen to her in the future, that she has very light symptoms now but that she could be in a wheelchair in a few years. She was a bit less sure about what she would want to do if the results showed that the baby had fewer repeats than she but realized that such an outcome was very unlikely. Peter reminded Marianne that they had a 50 percent chance that the baby didn't carry the gene at all, which, it turned out, was exactly what the tests eventually showed.

Insurance companies, the state, and taxpayers all have a financial interest in promoting prenatal testing for genetic disorders because of the expense of ongoing care required for many people with genetic conditions. This interest may have clinical consequences for people seeking information about issues other than their own reproduction at the clinic. Ad, a junior clinician, made this clear in his discussion of the range of reasons that motivate people to come to the clinical genetics centers:

> **KST:** What genetic information do you think it is most important for people to know?
>
> **Ad:** Mmmmm. I think it differs. I do not know if I can answer that or not because people come with so many different questions and different kinds of problems and that is the point that they want to know something about. So, very often when people come they have a certain problem in the family or maybe they themselves are the ones with the disease and they want to know more about the outcome which also can be dependent on their genetic makeup. And they also—when I offer the investigations I take blood from them and they fairly often ask, "You are only looking at that one thing, right? You are not messing with my other

genes, right?" It is very often that people say that. Now we assure them that we are only capable of looking at one thing at this time. . . . So people come with specific questions. And in that sense some people have often been afraid for a long time. . . . I think that it is only useful to try to answer questions that people already have. And sometimes, because of the way it works in Holland, people come with very specific questions, for example, a diagnosis, and we are . . . genetic departments are paid for giving, for not answering specific questions, but for giving what we call advice—*erfelijkheid voorlichting* [genetic counseling]—counseling, well, counseling, but we are paid to tell people what their risk will be that their offspring have diseases. But fairly often this is not the question that people have. Sometimes they want to know "what does this disease really mean? And how does it work, how does it work on a genetic level?" They want to know about this gene but sometimes they do not even want to have children anymore. But because of the situation, that we are paid to give people advice on the risks of their offspring, we often do not even listen to the questions that they really have, because we have to tell the whole story of "your children have a 25 percent risk and you can do prenatal investigations or you cannot do them." And so, people sometimes come for diagnosis and we force them through the whole circuit. And I think that is a bad situation. . . . We are paid to give predictions about the future, especially of offspring, but people often come to us for other reasons. They do not even want to have children. And, well, it is important to know [what people want to know] before you start talking to these people.

Lauren also spoke of the tension between the questions people have when they come to the clinic and the clinic's orientation around genetic counseling:

KST: What genetic information do you think it is most important for people to know and why?

Lauren: Well, that is an easy one. . . . I first have to find out what they want to know. Well, what I do find important for everyone to know, what I will tell them about in any case, is if it is about an inheritable disease and if there is a way to ensure healthy children. . . . But my responsibility stops once I have informed them: "This is what we found, the family may be at risk." It is their job to inform the family whether they want more family members tested.

Ad and Lauren both thought it was important to find out what kind of information people coming through the clinic desire. They understood that patients may not be seeking reproductive information. Ad raised specific concerns about the institutional economic incentives to obtain and deliver such information, while Lauren's comments suggest that she does provide such information to her patients. Janneke, a third junior clinician, expressed her desire to provide reproductive information rather than focusing only on the questions an individual or couple might bring to the clinic. She repeatedly told me that she was particularly concerned that people would be very unhappy to have a child with a genetic anomaly only to learn later that it was something they could have known about in advance.

A third factor physicians raised is that clinical diagnoses have consequences for scientific research because both clinical and research geneticists expect to find a common cause for shared symptoms. In this context, distinguishing subtle differences in morphogenic, metabolic, skeletal, muscular, mental, developmental, behavioral, or other qualities and characteristics becomes an aid in developing highly refined categories of people whose anomalies might have a common genetic explanation.

Fourth, clinicians see diagnosis as a source of emotional help to the affected persons and their families. Several physicians reported that they felt parents gained enormous relief from finding out that the combination of symptoms their child exhibited could be explained by a named syndrome shared by others.

Lauren touched on several of the factors described above while explaining the importance of accurate diagnosis:

> From a genetic point of view it is essential to make the right diagnosis because one thing could be autosomal recessive and the other thing dominant, or perhaps a mutation or anyway something that has no consequences for the family or the next pregnancy. That is why your answers, even if you have seen several children with that same syndrome, your answers . . . are not always the same. It is also about what the people want to know. There are a lot of people, parents, who come here who say, "I want to know what he has because this or that," but there are all sorts of reasons why they want to know what the child has and you can help them by labeling. And then . . . for example, if there is a syndrome that we know will prevent a child from speaking properly then I tell them, "Don't put your energy into speaking. I know it is

important for you that your child speaks but with this syndrome we know that he or she will not be able to speak ever. It is very tough but it is true. Put your energy into this child by teaching him how to express [himself] by [other] techniques . . . by helping him." And that is why it [diagnosis] is important because one thing that it [the condition] looks like may be children who do talk and that is very different and that is also why it is so important to have the [clinical] picture complete. And what I always stress to people is that we do not know everything about genetics; because we do not know we cannot do an exclusive test that says they have the syndrome or they do not. Probably [what we now call] a syndrome is not a syndrome. A lot of children who look alike are put together and maybe in ten years' time, or twenty, or one hundred, it does not matter, we can distinguish between them because look-alikes might be two syndromes and we can say in this category we have an A category and they have this DNA abnormality and we have a B category and they have the same features but they do not have the DNA abnormality . . . I say to [patients], "What I am telling you now, now we will say you have a child with an autosomal recessive disease with a 25 percent recurrence risk but maybe in twenty years' time we can say, "Your child has this disease but it appears to be a new mutation and you do not have this 25 percent recurrence risk."

Bart used the example of Rubenstein-Taybi syndrome when raising the subtle ambiguities involved in how clinicians communicate information to patients. In his discussion of these ambiguities he emphasized that in some cases people simply needed the reassurance that comes with diagnosis:

When they [clinicians] are not sure whether or not it is a specific syndrome . . . it forces the problem of "What are you going to say to these people?" Are you going to say, "You are very likely to have Rubenstein-Taybi," or are you going to say, "I think you could have Rubenstein-Taybi," or are you going to say, "You could have Rubenstein-Taybi?" . . . People are weighing these things in making decisions in their lives—based on what you tell them. But sometimes . . . when people are, for example, very much in need of a name of a diagnosis, because they have been looking for ten or more years for [the answer to the question], "What does my son have?" and you think it might be Rubenstein-Taybi, better to say, "Well, I think it is probably Rubenstein-Taybi" and not be so vague because these people need information. They need a diagnosis.

Other clinicians discussed the purpose and importance of accurate diagnosis with me, stressing the rarity of genetic disorders and subtle distinctions between similar disorders. Geert, a junior clinician, and Janneke talked about the confidence that comes from consultating with and finding agreement among one's colleagues:

Geert: The reason that we take the time with patients and why we spend so much time consulting with each other has to do with the rarity of the disorders as well. You can never get a lot of experience with things that have been seen only ten or twenty times in this world.
KST: Perhaps that is why one physician told me that clinical diagnoses are often based on a hunch?
Geert: Yes, and you feel quite relieved if someone agrees with you. Then you feel much better in talking to the people or to the physician that sent them here.

One day Janneke asked me what I thought about what I had been observing as I conducted my research at the clinic. At the time I was thinking quite a lot about the differences between laboratory and clinical diagnoses and the production of diagnoses in the audit. I explained to Janneke that one of the things I was interested in was how the clinicians developed proof of a diagnosis. I said that it seemed they drew on a combination of clinical and laboratory evidence to support diagnoses, that they seemed to need a certain amount of evidence to prove a particular diagnosis, but they did not always get that evidence from the same place—in some cases getting evidence from the laboratory was important and in other cases not. At one point in the conversation she picked up on my interest in developing diagnoses, saying,

Janneke: Well, if you are thinking about proof in that way then you also have to consider the audit and the LOG.
KST: I know. I think those are very important.
Janneke: Because if a lot of your colleagues agree with you then you are really sure about your proof. You are really sure you have it.

Geert and Janneke both accentuate the social relations involved in producing diagnoses, pointing out not only that they rely on their colleagues' expertise in producing diagnoses, but also that diagnoses become more reliable, more certain, when there is consensus among clinicians about them. Indeed, the more puzzling or complicated a case is, the more essential consultation and consensus become in developing a reliable diagnosis. The clinicians mobilize a wide array of evidence to convince their colleagues of the truth they are working to produce.

The physicians at the genetics centers are well aware of the social power their diagnoses have for affected individuals and their families. One senior clinician, discussing the fact that laboratory confirmation of clinical diagnosis is not always possible, said he thought it was important to be willing not to make a diagnosis. Hans, another senior physician, stressed that it was better not even to mention possible diagnoses to patients until one was sure about the case because once a physician names a potential diagnosis, it is difficult for patients to let go of that possibility. At the end of clinical appointments people would often ask whether the clinicians had any ideas about possible diagnoses. Hans, as well as other physicians whose appointments I sat in on, would say that he had some ideas but did not want to say anything until he had investigated them further. He explained this policy to me by telling me about the problems a family had experienced after being given a diagnosis that was later found to be inaccurate. Apparently the family had become quite involved with a support group for people with the disorder named in the first diagnosis, developing extensive social contacts based on it, and later experiencing great emotional difficulty with the disruption of their social network when the new diagnosis was produced. Janneke explained the importance of accurate diagnosis to the patient by noting the social power of diagnosis. She told me that "once you give a diagnosis to someone they live with that diagnosis, they are that diagnosis. I think you have to be very careful." The concern to be cautious in delivering diagnoses is striking in contrast to the intense desire to develop diagnoses among experts in the clinic.

INTERPRETING DIAGNOSTIC EVIDENCE

Clinical geneticists in the Netherlands initially see people at outpatient clinics in regional hospitals or at the clinic itself. During the first appointment, clinicians take a medical history and do a physical exam. The exam includes looking at the person (who has been asked to remove all their clothing except for their underwear), particularly at the face, head, hands, and feet, and measuring the length of the ears, the distance between the pupils, the circumference of the head, height, weight, and span, and the length from the tip of the middle finger to the beginning of the wrist. If possible, clinicians also take head circumference measurements of the family members present at the appointment.

The physical exam ends with the clinicians' request to take photos of the individual for whom a diagnosis is sought. The clinicians take photos of any obvious dysmorphic features an individual exhibits as well as of the person's body from all four sides, often including close-ups of the face and profile in addition to full-length shots, and of the hands and feet. They also take care to exhibit the ears in the profile shots. It is at these sites—the face, hands, feet, and ears—that the physicians expect they might find the subtle clues leading them to produce a diagnosis that allows them to categorize a difference already perceived in the person they are examining.

The geneticists often take a photograph of the patient with other family members. They may, for example, capture the face of a parent who helps with the photographing by holding the child they brought in for diagnosis or catch both parents with their child at the end of the photographing stage of an appointment. At the audit, clinicians frequently ask if there are any pictures of the parents, often framing it as "Are there any pictures of Dad?" or "Did you get any pictures of Mom?"[9]

At the end of the initial appointment the clinicians give the patients forms requesting information about the medical, educational, and employment background of members of the extended family and ask the patients to fill them out and return them to the clinic in a preaddressed envelope. These forms are quite extensive, asking for medical information for four generations. The patients are told that the geneticists will do some research—on computers and in consultation with colleagues—and that after the clinic receives the family's information the patients will be invited to make a follow-up appointment. The geneticist responsible for the case then compiles the family data as well as the information gathered at the first appointment and searches for a diagnosis, primarily through reliance on her or his training and experience, computer searches, and consultation with colleagues. If a geneticist finds that the combination of symptoms in a particular patient fails to indicate a clear diagnosis he or she then discusses the case with a colleague or two. If this consultation does not resolve the ambiguity, he or she then brings the case to all colleagues at the audit.

I was struck by the use of photographs in the presentation of patients at the audit. Their ubiquity directs attention to the power and significance of visual images. John Berger (1973) argues that multiple meanings inhere in visual images and that their interpretation involves political and aesthetic judgments. Building on Berger, Peter Jackson draws parallels between political representation and visual representation. Jackson points to the power of visual images when he argues that "for any claim by one group to 'represent' another is itself a form of power, exercised over subordinate groups by those more powerful than themselves" (Jackson 1992:89). The ability to impose one's reading on a visual image is an act of power.

In a medical context, the historian Ludmilla Jordanova argues, the very familiarity of photographs can invest them with special authority. She observes that "the photographs now so often used as illustrations in medical works create a dramatic impact by their unrelenting literalism, especially if they are in colour" (Jordanova 1989:140). In the audit, I was sometimes astonished when I thought I was looking at photos of a perfectly normal child only to have the clinicians say that the face was coarse or comment on the hairline or color, the shape of the nose, the placement of the ears, the length of the philtrum, or the width of the forehead. Looking at other photos, many of which presented severely dysmorphic individuals or fetuses, I was disturbed by how much can go wrong during development. These pictures showed a broad spectrum of people, both male and female, from many social backgrounds and of all ages and occasionally included photos of fetuses from terminated pregnancies.[10]

In his work on the production of visual images of brains, Joseph Dumit points out that the power of visual images is frequently stated but rarely explained. He argues that the power of images is culturally and historically specific, having to do with the power and value attributed to scientific practices. Such images gain their authority from having been produced by science (Dumit 2003). But the images Dumit deals with are highly digitalized positron emission tomography (PET) brain scans, which require an expert to be interpreted. In such a situation the scientific aspect of the images is doubled—they are produced by technology developed out of scientific knowledge and they are considered readable only by an individual familiar with the scientific knowledge supporting the technology.

Photographs of people's bodies do not automatically represent scientific knowledge in the way that Dumit's digitalized images do. In the audit, the clinicians read science and scientific knowledge into the photographs. It is in the act of reading the photograph, which is also an act of reading the body and, by extension, of reading the genes, that science inheres in the photographs. Through the process of interpreting the photographs, seeing and knowing are conflated and scientific knowledge is produced.

Although the clinicians consider the photos an important component of producing diagnoses, they acknowledge that their focus on visual representations of subtle dysmorphic features is problematic. As Hans jokingly put it, "Everybody has a syndrome. I have a syndrome. I have the same syndrome my father had," suggesting that one could find dysmorphic features in any person.

Another example brings into relief both the questionable reliability of photos and assumptions about the objectivity or scientific nature of photos. In this case, Bart explained that he had a patient, an adult man who was mentally retarded, who insistently refused to have photos taken of him during an appointment with a geneticist. The patient's father, a professional photographer, asked the clinician why he wanted the photos and then offered to take some himself and send them to the clinic. When Bart presented this case at the audit he passed the photos around, explaining the circumstances under which they were taken, emphasizing that the father is a professional photographer and commenting that "the pictures are very beautiful and do not show everything that should be discussed in this person." From Bart's point of view there were two problems with the photos. First, they did not illustrate the dysmorphic features he had observed in his examination of the patient, and, second, they did not show the level of dysmorphism that the clinicians would expect to see in a person with the symptoms Bart described. Bart is distinguishing between the aesthetic and personal value of photos as sentimental representations of loved ones versus the scientific value of photographs produced to serve clinical ends. To support his diagnosis, he must segregate the personal aspects of representation in order to reinforce the authority of the photos' scientific value.

The discussion following Bart's presentation demonstrates the clinicians' struggle to correlate the symptoms described with the morphogenic features they saw in the "beautiful pictures," which did not represent or reflect those symptoms:

Sabine: How severe is the mental retardation?

Lauren: Why does he have scars on his stomach?

Bart: The scars on his stomach are because he stabbed himself because he thought he was pregnant because he is fat and he thought the baby needed to be taken out.

Lauren: So, he is not very normal.

Bart: No, he is retarded.

This case helps illuminate the reliance clinicians place on photos produced in a clinical or scientific setting and their belief that such images capture something they perceive as objective that they can later utilize as a means of explanation and interpretation. It further demonstrates the power of the images by illuminating the cognitive dissonance that occurs among the clinicians when photos do not reflect the data in a case.

The normalizing process taking place in the audit actually has a double function: at the same time that it serves to normalize the bodies of patients, it normalizes the bodies of the physicians themselves. For example, the senior staff actively encourage the residents to present cases at the audit as well as at the monthly national meeting of clinical geneticists from all eight of the genetics centers, a portion of which precisely reproduces the presentation of patients in the audit. They are expected to present a minimum number of such cases during their training.

The physicians' experience of the normalizing aspect of their work is perhaps most clearly articulated in an incident that occurred just before my arrival in the Netherlands. In this case two junior clinicians jestingly challenged the diagnostic process by presenting photos of themselves as children at the audit, representing the faux cases as patients with slight mental retardation. Their colleagues began discussing the dysmorphic features they saw in the pictures. The two junior clinicians found this rather unremarkable because there is general acknowledgment among the clinicians that everybody is dysmorphic in some way. But, they told me, they were surprised at the number of things their colleagues found to discuss and the length of time and number of hints it took for them to catch on to the prank. While one senior geneticist was reportedly quite displeased by this event, another discussed the incident with me, saying, "It just shows how important it is to be cautious when using photographs." This incident indicates that seeing and interpreting bodies in the way physicians do in the audit is not something that comes easily or that everybody is comfortable with (at least when one sees it applied to oneself), but rather is something developed through the professionalization of young geneticists.

The clinicians consider the ability to read photographs for diagnostic purposes a skill that varies from person to person. Their training requires them to regularly present patients in order to develop their skills in this area. The presentation of patients' photos also serves to familiarize clinicians with variability within diagnostic categories. Janneke told me that one of the purposes of using the photographs in the audit is to help the physicians "learn more about the spectrum of a syndrome." Bart and Saskia, a junior physician only recently beginning her training in clinical genetics, were less than confident about the reliability of photographic evidence. Saskia explained, for example, that she prefers dealing with skeletal cases because she finds diagnosing them "more objective" than diagnosing other types of cases. For Saskia the objectivity in such cases stems from the concreteness she perceives in X-rays and skeletal measurements. Bart told me frankly that he does not think he is very good at interpreting photographs of patients. In my interview with him I described my reactions to the photographs in the audit and asked him about his reactions:

> **KST:** One of the things I am really curious about is the use of photographs in the meetings. Perhaps because I'm not trained as a doctor sometimes they really shock me.
> **Bart:** Some of the pictures are really personal.
> **KST:** Yes. Sometimes I am surprised because they just seem so personal and sometimes because I think, "There is nothing wrong with that child [so] why are we looking at this picture."
> **Bart:** Hm-hm. That is what I think very often.
> **KST:** Other times I think, "How could so many things go wrong?" [Laughter] I have very different reactions to them so I guess I am curious about your reactions to them.
> **Bart:** Well, I do not tend to look at them from a personal point of view. I have the same reaction that you often have: That there is nothing wrong with this child. And then people [the other physicians] start to come up with all these dysmorphic features and I think, "Well, okay, I have that [dysmorphic feature] too." [Laughter] But no, they do not shock me anymore. They are not meant to be looked at from a personal point of view because if they were for personal use, they would not be taken. They are only taken to convey information to other colleagues, and it is also easier to talk about a child when you know what he looks like, especially in these cases. . . . We always ask for permission, and I always explain to people what I want the photographs for—because I am not

sure [of a diagnosis] myself, and I want to discuss it with others and it is easier if we have a picture to look at.

In spite of their acknowledgment of the problems associated with using photos in the way I have described, the physicians consider them an essential component of presenting a case for diagnosis. I witnessed only one case in which the geneticist in charge of it had chosen not to take photos of the patient in question. On the basis of the patient's medical history the clinician suspected a particular diagnosis associated with severely dysmorphic features. The clinician in charge of the case, however, thought the patient "looks so normal" that he decided not to take any photos. When he presented the case at the audit several of his colleagues asked about pictures. One senior geneticist said they needed to see pictures, and another geneticist asked, "But would you make such a diagnosis in someone who looks completely normal?" Another physician commented, "In the books the pictures they show with this syndrome are always of people who look very strange." Another colleague concurred, saying, "I agree . . . that making such a diagnosis of a normal looking person is a problem because the pictures you see of these people are always very strange." Still another senior physician said, "Yes, but they always use the most extreme cases in the books." The presenting geneticist repeated his question about whether anyone would make this diagnosis in a person that "looked completely normal," and the senior geneticist who had initially said they needed photos said, "Yes, I would, if I saw the person myself. Otherwise you have [to have] the photographs."

BETWEEN CLINIC AND LABORATORY

In addition to photographs, geneticists produce and use other images of the body as they seek to make sense of genetic anomalies. These visual representations reflect the various genetic foci of the center's different departments. Those carrying out research or diagnostic tests at the DNA level represent the body in photographic images of DNA on laboratory gels. The cytogenetic department staff illustrate the body with highly magnified pictures of whole chromosomes, typically presented in karyotypes or stained through FISH.[11] The clinicians rely on these representations from the center's laboratories as well as on representations of the body derived from diagnostic tests conducted by other medical specialties (e.g., X-rays, CAT-scans, and metabolic test results) at the same time that they use photographic representations of patients to interpret bodies in terms of genetics. Geneticists interpret these various images as containing scientific and medical truths about the body.

Competing efforts are made to impose definitive readings on various representations of the body rendered in terms of genetics through images and readings produced in the clinic and in the laboratory. At the genetics centers various types of visual representations are differently valued and ranked hierarchically: the most technologically enhanced representations, those most in need of technoscientific amplification and expert interpretation, are the most highly valued and seen as most accurately representing truth about the body. In her call for scientific practices grounded in situated knowledge, Donna Haraway cites the power of vision in scientific practice to create a "politics of positioning" such that all knowledge is situated (Haraway 1991:193). In the world of genetic research, where technological enhancement of images is the sine qua non of practices both in the laboratory and the clinic, optics as a politics of positioning pervades the visualization of bodies. Haraway also points out that "the 'eyes' made available in modern technological sciences shatter any idea of passive vision" (Haraway 1991:190); in other words, all vision in the contemporary biosciences is actively mediated.

Jordanova emphasizes that vision has long been a central component of the way scientific and medical truths are produced: "The process of looking is central to the acquisition of valid knowledge of nature. From classical times, science and medicine have been explicitly concerned with the correct interpretation of visual signs, and skills in those fields was preeminently seen as a form of visual acuteness" (Jordanova 1989:91). Those who establish their reading of a body as definitive gain the power to impose its meaning and implications both upon those whom the images represent and those who situate themselves in reference to the image. Nevertheless, in spite of their power and their hierarchical ranking among medical and scientific professionals, the meaning and value of these various representations are contested in the daily practices within and outside of the clinic. Thus, readings of the body in relation to genetic practices involve contests of power not only over the subject being represented, but also among laboratory and clinical geneticists themselves as they vie to establish definitive diagnoses in the eyes of their colleagues and their patients.

The process of diagnosing genetic disorders involves negotiation between the clinic and the cytogenetic and DNA diagnostic laboratories with which it works. Earlier in this chapter I explored the dynamic processes in the audit through which clinical diagnoses are achieved when there are no definitive laboratory tests available. In practice, laboratory diagnosis most often provides confirmation for a suspected clinical diagnosis. Both laboratory and clinical geneticists generally consider laboratory tests definitive. In cases in which a syndrome is well known and relatively common and the patients' symptoms are strikingly characteristic of that syndrome, physicians find clinical diagnosis a straightforward affair. This is often the case, for example, with Down syndrome, for which the morphogenic characteristics of the body and of the chromosomes are well known, and with alkaptonuria, for which the primary symptom, black urine, is unmistakable.

The relative simplicity of these types of cases is evident in geneticists' attitudes about laboratory diagnoses in general and in their attitudes toward diagnoses developed through DNA analysis in particular. In discussing with me the value attached to laboratory diagnosis, Hans stressed the stability of knowledge developed through DNA testing:

> **KST:** My next question is, well, there are cases where you are quite comfortable making clinical diagnoses based on experience. But, it still seems like you are most comfortable, and happiest, if you get chromosomes [that confirm the diagnosis].
>
> **Hans:** Yes, or DNA.
>
> **KST:** Or DNA. Chromosomes, FISH, or DNA. That seems to be what you consider final proof. . . .
>
> **Hans:** Yes . . .
>
> **KST:** What is it about getting the proof from the DNA that makes you so confident?
>
> **Hans:** Yeah, yeah, yeah. And are we looking to the real proof? Because you have in chromosomes and FISH, you see the proof. You see there is something missing. In DNA, you see it, you see the abnormality. You think . . . that it is objective. Not subjective. I think that gives you the confident feeling. You can show it to other people—"See, I am right. Here is the proof." It is not necessary to believe or not to believe [in its existence]. No, it is. You can get a figure. You can list it, you can count it, etc., etc. And, it is not anymore those strange dysmorphologists who say, "Oh, this syndrome, that syndrome." One strange syndrome after

another. No, it is real. It exists. [You can say], "Look, there is something abnormal [in the DNA]."

Hans stresses the importance of being able to visualize genetic anomalies at the level of chromosomes and DNA as provided through laboratory analysis. His comments point to a hierarchy of representation that privileges those images of the body produced by the greatest degree of technological intervention and most thoroughly interpreted through genetics. Hans's sentiments were echoed by Meik, a junior clinician. When I asked Meik about the relationship between the clinic and laboratories in the production of diagnoses he explained that laboratory confirmation boosts the clinicians' confidence about clinical diagnoses.

In another conversation, Pim, a molecular biologist, reiterated the prevailing hierarchy of representations of the genetic body. I suggested that perhaps molecular genetics, involving DNA gels (fig. 2), was more abstract than cytogenetics, involving karyotypes (fig. 3) and FISH (fig. 4). He became quite defensive, saying, "It is not abstract, not abstract at all. One of the nice things about molecular genetics is that it is not abstract." He asserted that at the DNA level, geneticists are "looking at a deviation of a normal pattern and that is it. We are looking for one base that is different." Pim points to the importance of normalizing the body at the level of both phenotype and genotype. He contrasted his work in DNA labs with the work of cytogeneticists, who look at whole chromosomes. He said that cytogeneticists are "looking at too large a scale; cytogenetics is more abstract." This molecular geneticist was contrasting what he considers the concrete nature of molecular genetics, which looks at DNA at the level of its constituent bases, to the abstract interpretations he believes cytogeneticists may make in describing whole chromosomes. He sees cytogeneticists' interpretations as being more abstract because the representations on which they are based are of whole chromosomes in which the constituent bases of DNA cannot be seen. Thus, chromosome analysis may be able to identify a difference in a chromosome, but it may not be able to explain precisely what the specific difference involves. For these geneticists—clinicians and laboratory scientists—truth about the body appears to lie at the molecular level. The closer an image comes to representing this level, the greater the authority it carries.

The role of images in the clinic exemplifies Jordanova's observation that "Western scientific traditions have customarily placed 'looking' at centre

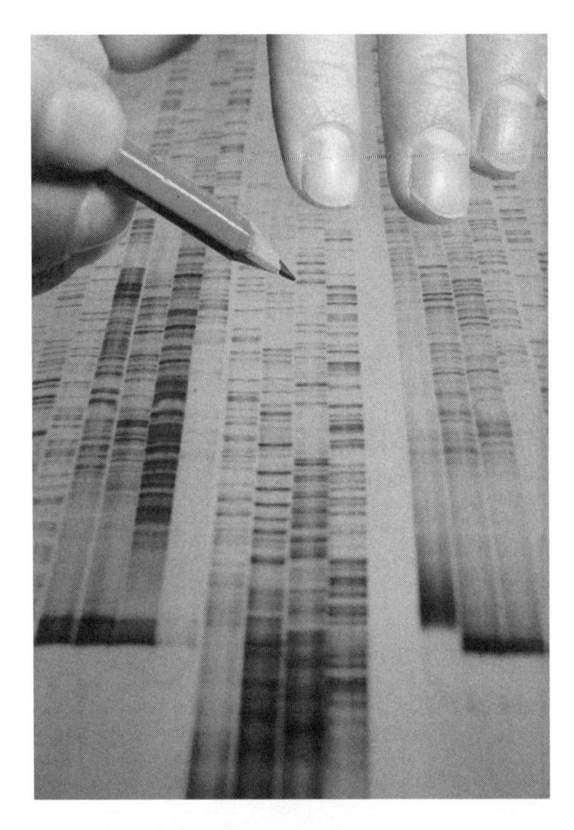

2. DNA gel.

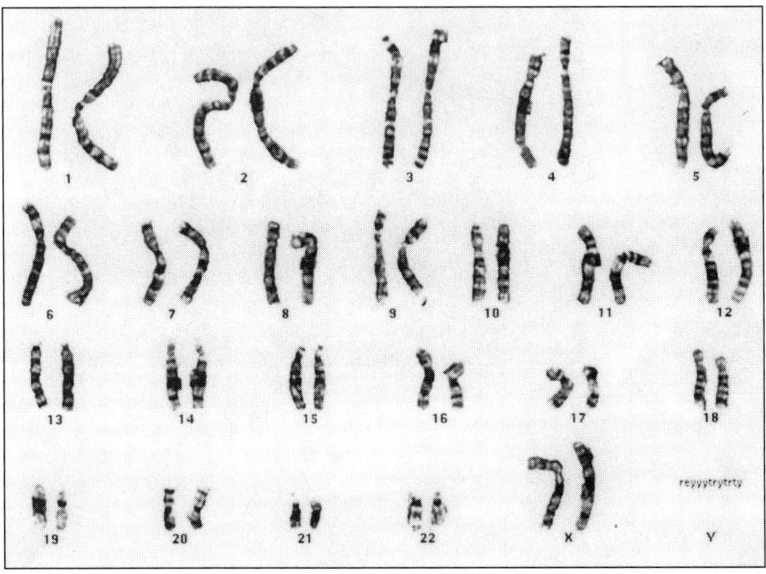

3. Karyotype.

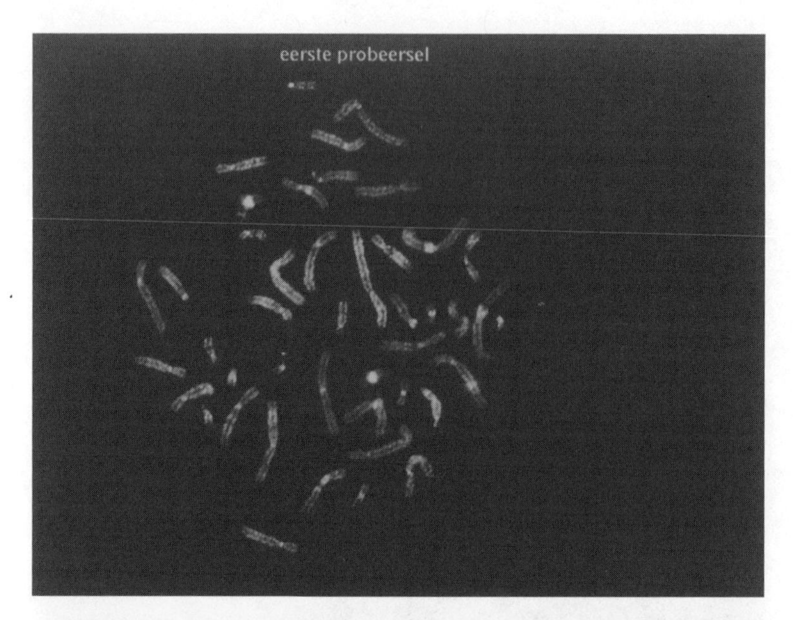

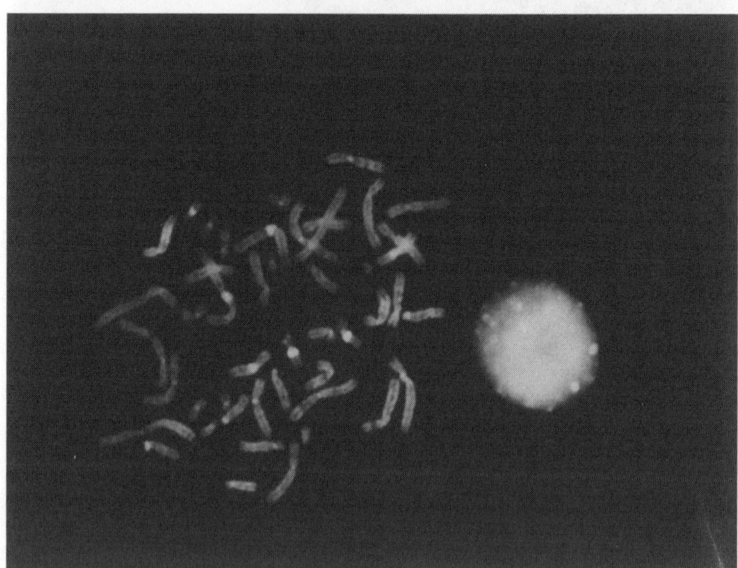

4. FISH (Fluorescent In Situ Hybridization).

stage, although, to be sure, they have done so in historically specific ways" (Jordanova 1989:91). In discussing such nineteenth-century medical representations of human anatomy, Jordanova suggests that the central function of physiognomic traditions was relating the face or form to character or inner qualities, "moving inferentially from visual signifiers to other, invisible, inner signifieds." She writes, "Physiognomy, as an exceptionally pervasive historical phenomenon, offers one clue as to why scientific knowledge was associated with vision, and why its success was linked with removing the impediments to vision" (Jordanova 1989:92). Jordanova's observations resonate with the privileged status contemporary geneticists accord to molecular images of the body represented in terms of genetics. Geneticists embrace molecular representations of DNA as overcoming all impediments to vision. Even more than chromosome karyotypes, DNA gels appear to represent an ultimate, previously invisible inner truth.

In spite of widespread medical, scientific, and popular understandings of the body as biologically given, the idea that the body, including its anatomy and physiology, is socially and historically produced is now well understood (e.g., Duden 1991; Kuriyama 1999; Langford 2002; Lock 1993, 2001; Martin 1987, 1994; Morgan 1989). In Enlightenment traditions since at least the eighteenth century scientific discourses have represented the body as a "natural phenomenon, while at the same time it has been made invisible as a social creation" (Duden 1991:20).

In molecular representations of the body, geneticists' identification of truth with science is so complete that they do not seem aware of the degree to which these images themselves signify the power and history of scientific knowledge. Authority here derives from the identification of science with truth—the pictures are not explicitly understood as having been produced by science so much as they are considered to represent science's unveiling of nature's preexisting truth about the body. There is, however, still a process of signification going on, most obviously in medical and scientific interpretations of the images, which require expert training and knowledge.

There is an irony to medical and scientific interpretations of laboratory images as transparent representations of natural truth about the body in that these images are themselves highly manipulated. In producing karyotypes, for example, laboratory technicians stain, magnify, and completely rearrange chromosomes into pairings that do not occur naturally (fig. 5a). Most of the time individual chromosomes do not look the way they do in their representation in karyotypes but appear as indistinct threads within the nucleus of a cell (fig. 5b, upper right-hand corner). To produce usable images laboratory technicians must catch the cell just before it divides, that is, in metaphase, because this is when it can be made most visible. An image produced in this manner and magnified as many as two thousand times or more shows chromosomes in a disorderly jumble (fig. 5b). To create the paired karyotype, laboratory technicians take individual chromosomes out of this jumbled context and rearrange them according to size and banding patterns (produced by staining) into numbered and ordered pairs. Laboratory geneticists cut and paste these images (either by hand or through computer technology) into new arrangements in order to get at what they interpret as a deeper scientific and medical truth about the genetic body.

The geneticists' unself-conscious identification of the karyotype or DNA gel with objective reality tends to efface the mediating role of technology and human agency in producing the image. By taking technology for granted, geneticists overlook the degree to which they themselves assume the aspect of a cyborg—an entity in which the boundaries between human and machine are blurred. The geneticists identify their vision with the vision of the image-producing technologies and machines in the laboratories. They speak of seeing a gene, not of seeing a technologically enhanced image of a gene. The technologies and machines that produce the images thereby implicitly become the geneticists' eyes.

Molecular images provide authority to represent truth about the body and confidence in dealing with patients. Hans and Meik both explicitly refer to the confidence they gain when one of their clinical diagnoses is confirmed by laboratory tests. Hans also refers to the images to bolster patients' confidence in his diagnosis and by extension in his expert authority. To the degree that patients accept laboratory images as representations of a scientific or medical truth about their bodies, they accede to the authority of somebody (the clinician) whom they perceive as able to interpret them. Additionally, a clinician's mastery of images promotes confidence among her colleagues in her abilities and status.

In the absence of laboratory confirmation, clinical diagnoses are often considered ambiguous and open to further interpretation. Ambiguity is most evident in cases like those presented at the audit and the LOG, in which patients exhibit symptoms that do not point to any obvious diagnosis. The activities associated with clinical diagnosis that I described above in the audit and the LOG shed light on the work that goes into

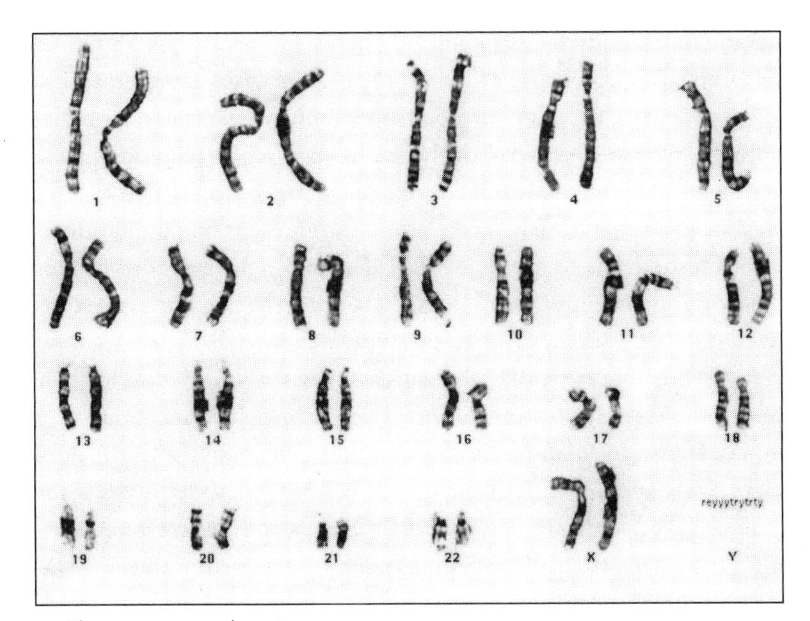

5a. Chromosomes as karyotype.

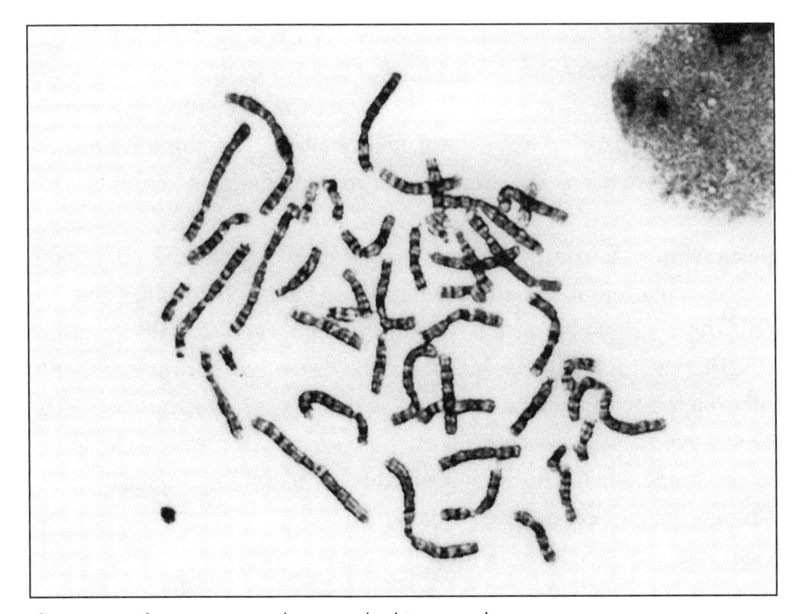

5b. Human chromosomes photographed in metaphase.

ameliorating ambiguity by distinguishing between normal and abnormal as well as between Dutch and other in the puzzling cases. Laboratory and clinical geneticists alike often interpret the problems associated with distinguishing such categories as largely arising from not having found the specific genes in the genome responsible for many disorders that are believed to have a genetic basis. Their attitudes represent one limitation of laboratory testing: There are many disorders and named syndromes for which laboratory diagnosis is not possible because the genes that are considered the source of many disorders are not known.

Even when laboratory tests are available they may not be exhaustive. While certain tests may provide definitive positive identifications of a syndrome, failure to identify a syndrome does not always mean it does not exist. For example, at the time of my fieldwork a laboratory test for Rubenstein-Taybi syndrome was available. The test, however, offered definitive diagnosis of a particular deletion in only 20 percent of the cases labeled with a Rubenstein-Taybi diagnosis. The uncertainty inherent in such a situation was raised in an audit discussion about a possible Rubenstein-Taybi diagnosis. In this case the clinician in charge suspected that one of her patients had Rubenstein-Taybi and was inquiring about the possibility of laboratory confirmation for the diagnosis. In the discussion of the case, one of the geneticists from the cytogenetics laboratory pointed out that "Rubenstein-Taybi is a clinical diagnosis. You find a deletion in only 20 percent of the cases."

The indefinite nature of clinical diagnoses is clear in the comments from Lauren quoted above about what constitutes a syndrome, and the fact that there simply is not a laboratory test for every condition. In acknowledging the limits of current diagnostic processes, Lauren highlights the open-ended nature of the knowledge contemporary geneticists produce. In this conception of syndromes, the syndrome serves as a classificatory place-holder—allowing the possibility of temporary normalization until something more definitive comes along.

Despite the limitations of laboratory tests, clinicians did not question the status of the tests themselves as representations of reality. Indeed, some clinicians appear to cope with uncertainty by asserting that any limitations in laboratory testing ultimately can be overcome through further application and refinement of scientific knowledge about the human genome. Geert discussed the reasons for his uncertainty about laboratory tests by saying, "I am not so worried about it [the current limits of laboratory tests]. I think it is just a matter of time. . . . I think some day they may find other [deletions in the genome] with which you get a sharper delineation of the clinical picture." In contrast to Geert's faith in the ultimate reach of laboratory tests, Hans expressed a belief that testing would never fully supplant the interpretive judgment of the clinician in diagnosing genetic disorders. In so doing he contrasts different forms of authority:

> It will always be necessary that the clinician, a doctor, looks for certain symptoms. To weigh them—is this important or not important?—and include it or not [in developing the diagnosis]. . . . I think that a machine will never be able to do that. Yes, to a certain level. . . . I think to a certain level computers can do a lot but there will always be, it will always be necessary to weigh the symptoms . . . you will always have the clinical dysmorphology. I think, because people make strange combinations. People from Tunisia have children with people from Ireland. Is that child strange or not? Is it abnormal or not? A machine cannot give that answer. So there will [always] be work for me.

Hans asserts his authority by invoking clinical expertise in reading social categories of race and ethnicity that cannot be interpreted at the molecular level. In so doing he challenges the authority of the laboratory to bolster his own ability to produce truth.

In discussing the relationship between laboratory and clinic with me, Bart pointed to the ambiguity he considers to be inherent in clinical diagnosis:

> Some people are really good at diagnosis from dysmorphic features so they judge somebody's appearance . . . their attention is focused on certain dysmorphic features, things like thumbs or [other] things like that. Dysmorphic signs can point to one disease very strongly, but most of the time you are in a situation where you say, "Well, there are some dysmorphic features and it could be Rubenstein-Taybi [syndrome]." You are not completely sure, that is why you need your laboratory investigations. But for most syndromes, what I mean are dysmorphic syndromes, where there is only the dysmorphology that can help you, there is no laboratory confirmation available. So when these people [other clinicians] jump right up and look at the textbooks [during the audit], I have no problems with it. But when you are not sure whether or

> not it is a specific syndrome . . . they [the patient] could have something else . . . it forces the problem of what are you going to say to these people [the patients]? . . . That is the problem with dysmorphology, and I do not think I am very good at dysmorphology. I would like to be able to depend more on laboratory investigations because people will make . . . at least we think, that they [laboratory tests] are right, that they are exclusive. People [patients] are weighing these things in making decisions in their lives—based on what you tell them.

Bart's comments indicate the indeterminate nature of many clinical encounters. He implicitly invokes the hierarchy of value associated with laboratory representations of the body by appealing to laboratory tests for diagnostic confirmation. Yet, even as he does so, Bart qualifies their authority with the words, "at least we think that they are right." This gesture suggests that even representations such as karyotypes may remain open to further interpretation.

Junior clinicians tended to have greater confidence than senior clinicians in the primacy of laboratory images as representations of truth about the genetic body. There was a greater range of attitudes about the status and value of laboratory images among senior clinicians. Junior clinicians may feel more comfortable relying on laboratory test results because they are still relatively inexperienced with clinical genetics practices. Their comfort with laboratory test results may also relate to the more formal training they now receive in genetics. Unlike their senior colleagues, many of whom came to genetics through a variety of medical specialties, those doing their residencies today do rotations in both the cytogenetic and the DNA laboratories. During these rotations they receive formal training in the laboratory techniques and interpretations associated with the various diagnostic methods. This formal introduction into laboratory techniques, which engages the minds and bodies of the residents, combined with the relatively short period of time spent in the clinic, may facilitate junior clinicians' faith in and reliance on laboratory test results.

Even among those senior clinicians who seem willing to contest the validity of laboratory test results, their challenges tend to focus on the particular images produced by the tests in specific cases rather than on the tests themselves or on the science and technology on which the tests rest. In discussing the relationship between clinical and laboratory diagnosis, Hans highlighted the situated nature of diagnosis:

> I have at least two children . . . with a fetal cardio-facial syndrome, and they do not have the . . . micro-deletion. I told them [the laboratory geneticists], "Look better. I do not believe it." For I think it is the fetal cardio-facial syndrome and . . . some . . . other people agree with me. It is not very scientific but it is a feeling based on, of course, more or less scientific listing and pluses and minuses, but also on experience. A child does not look like this or that. You can write down symptoms of children and you can make a diagnosis. . . . The description [of the syndrome] fits [the patient]. . . . Maaike [another senior clinician] has had many of those situations where she says, "It must be this, it must be that." For example, in hypotelorism. And then she looks at the chromosomes and then she goes back to the lab and says, "Look again. Make them [the chromosomes] longer. Stretch it, do it again, for they must have something." And I think that is because of the fact that you have a strong clinical feeling. . . . I think in making a clinical diagnosis you always have to be aware that you are making clinical diagnosis. That you do not have, for example, DNA or chromosomes, or metabolic evidence . . . it is a clinical diagnosis, and that remains until DNA [testing] is possible.

Hans's comments indicate that laboratory outcomes are not always automatically accepted by the medical staff. His approach to the two children in the above example highlights the fact that developing a diagnosis is a process of molding facts to fit a clinical diagnosis. Here, where Hans finds facts in the form of laboratory test results that do not fit his diagnosis he asks the laboratory to do the tests again. In this example, Hans uses his status as a clinical expert to challenge the value and meaning of laboratory images. Rather than automatically accept a laboratory outcome that does not support his clinical analysis, he questions the quality of the laboratory work that produced the test results. Thus in certain cases a diagnosis may be negotiated by repeating tests (or, as the work of the audit reveals, excluding symptoms) that do not support a particular clinical diagnosis.

Even in those cases in which laboratory tests are both available and exhaustive, they are sometimes subject to multiple interpretations. Different interpretations often reflect the different knowledge and experience of the various geneticists involved in a case. In these cases the evidence is made to say different things by two competing and contradictory communities of experts, who then must negotiate and contest the meaning of the images on which diagnosis rests. Such negotiations, while collegial, are also potentially dangerous. In opening up supposedly definitive results to multiple interpretations, geneticists risk undermining the entire range of scientific and medical authority through which the images were produced.

Certain contrasts between clinical and laboratory approaches to using and interpreting laboratory images are captured in a case involving the coincidence of a rare genetic condition in two unrelated individuals from the same street in the same town. At one morning meeting of the clinical staff, cytogeneticists, and the head of the DNA laboratory, a cytogeneticist presented a case of a chromosome anomaly known as a partial trisomy. The presenting geneticist explained that upon diagnosing this partial trisomy he had, like any good researcher, gone to his computer to search for similar cases. He described his surprise at finding a virtually identical case in the computer, his even greater surprise at discovering that this second case had been diagnosed at his own genetics center, and his astonishment when he learned that the two cases involved patients living on the same street in a medium-sized town not far from the center. Those present at the meeting also expressed astonishment at the coincidence of these two cases of the same extremely rare condition.

The reactions among those present went beyond astonishment, however. They engaged in some questioning, exchanged skeptical looks, and asked some amused veiled questions about the possibility that the two children involved in these cases might have the same father—that one of the children might be the result of an affair. I had already been told several times by physicians and laboratory geneticists that the center occasionally finds cases of what they call nonpaternity. What they mean by this is cases in which chromosome or DNA analysis reveals that a child does not have any chromosomes from the person that has been presented as the father.

After the presentation of the case I describe here, there was an element of titillation in the tone of the discussion. One of the clinicians lived in the same town as the patients involved in the case. Somebody at the meeting jokingly asked her, "Hey, what is going on in that town?" The clinic director immediately chastised the group for their improper attitudes. In a more somber manner, several laboratory geneticists wondered whether they should analyze the chromosomes for common paternity. One laboratory geneticist even turned to me and jovially said, "It would be interesting." As soon as the clinicians realized there would be no medical purpose to such tests, they expressed their strong disapproval of such a pursuit. Indeed, their attitudes indicated that the laboratory geneticists' suggestions amounted to an affront to their sense of medical and scientific propriety. This was made clear to me in a later conversation I had with the clinicians responsible for the patients involved in the cases the cytogeneticist had presented. When I spoke with them one told me that they were very upset about the joking, saying, "These are our patients. We know them." Here, again, there is a contest not only over what evidence means, but also over what it should be allowed to say.

This story illustrates how laboratory geneticists may extrapolate from the decontextualized images they produce to construct a potential social reality. Their approach in such cases conflicts with the social experiences of their patients. The clinicians' rejection of the laboratory geneticists' hypothesized reality echoes Hans's rejection of test results that did not conform with his clinical diagnosis. In both instances the clinicians use their situated knowledge to challenge certain aspects of laboratory geneticists' interpretation of an image. In the second case they are using their situated knowledge to resist a perceived incursion into their territory by the laboratory geneticists. The implicit message is that laboratory expertise should begin and end with the production and interpretation of the decontextualized image represented in DNA gels and karyotypes. The clinicians perceived the laboratory geneticists' venture into the realm of interpreting the social lives of patients as an improper transgression of the boundaries between clinic and laboratory. They objected because the laboratory geneticists' interest appeared to be driven by a combination of their scientific curiosity and titillation abstracted from concern for social and ethical implications.

The adjudication of meanings is not without constraints. Like the processes they represent, clinical and laboratory results are strongly interrelated. In questioning test results, Hans opens the door to skepticism about the deeper networks of scientific and technological knowledge upon which they are based. To avoid that potentiality, Hans limits his questioning to the quality of the results rather than to the scientific and technological principles that underlie them. In other words, he questioned the human technicians, not their techniques. He similarly challenged only one test result, not the ability of tests themselves to render definitive results. This strategy allows Hans to use clinical knowledge to negotiate the meaning of particular test results without threatening to undermine the deeper networks of science and technology in which they are embedded.

The genetics centers are made up of a constellation of intersecting domains that approach genetics from various positions. The people working within and negotiating across these various domains are deeply invested in various aspects of the practice of human genetics. A hierarchy of diagnostic proof operates in the genetics centers, where representations of the body gain more authority as they approach the molecular level. It is these molecular images, as represented in DNA gels, that geneticists see as most clearly representing truth about the body. This hierarchy, however, is not absolute. As Hans's difficulty with the case of the two children with fetal cardio-facial syndrome shows, clinicians, above all, senior clinicians, are willing to challenge certain test results based on a viewpoint situated by their clinical expertise. Yet even these challenges remain bounded and contained within the larger framework situated within prevailing paradigms of contemporary scientific and medical knowledge of genetics.

MAKING THE ANOMALOUS ORDINARY

The intense authority of visual images is a vital component of resolving the ambiguities presented by genetic anomalies. In the audit, the primary ambiguity requiring resolution is that of the distinction between normal and pathological (Canguilhem 1991 [1966]). The problem of Ehlers-Danlo syndrome, a dominantly inherited connective tissue disorder characterized by loose ligaments and often found in circus acrobats, underscores the ambiguity inherent in the categories of normal and abnormal. Whenever a suspected case of Ehlers-Danlo came up at the clinic Hans would groan. During an interview with him I asked him about his reaction to this syndrome. He told me that the syndrome is difficult. When I asked him why, he said it is difficult because "you do not have . . . you have nothing. In fact you have only eyes, ears, and feelings [to go on in making a diagnosis] . . . so you, I, find it very difficult to make a diagnosis. Because also there is a large overlap with so-called normal people . . . who is normal? I don't know. I think sometimes there is such a large overlap." Hans's comments are a testament to the power of clinicians to transform contested phenotypic evidence into clinically specified syndromes, particularly in the absence of molecular evidence produced in laboratories. Even as clinicians engage in this process, they articulate an awareness of the contingency of the categories they are producing.

Making distinctions between normal and abnormal can also take on a racial, ethnic, or national cast. National, ethnic, or racial background can become significant in developing genetic diagnoses because there are genetic conditions that have a higher incidence among or are specifically associated with distinct national, ethnic, or racial groups.[12] These include Tay-Sachs, sickle cell anemia, and thalassemia, which, often rather problematically, are associated, respectively, with people of Ashkenazi-Jewish and French-Canadian, West African, and Mediterranean origin and their descendants around the world.[13] These types of cases, though unusual, occasionally do arise at Dutch genetics centers. During my fieldwork, physicians at the clinic dealt with cases involving both sickle cell anemia and Tay-Sachs. Beyond these well-known conditions that are commonly associated with race and ethnicity, there were other clinical presentations at the genetics center in which race and ethnicity entered more subtly into the discussion. In these cases clinicians invoked nationally marked morphologies as an aid to constructing and interpreting genetic variation. These cases, rather than articulating differences in terms of normal and abnormal, implicated complex distinctions between Dutch and other.

At one audit Iris presented the case of Ali, a six-year-old Turkish boy. Ali had experienced numerous medical problems that began two hours after birth, and during his lifetime he had already seen numerous physicians for his various problems. When his case was presented at the audit we were told that, among other things, his ears are low set, he has a flat nose and philtrum, his uvula is small and his lips thin, his neck is short, his hairline is low on the back of his neck, and his penis is small. After the presentation of these problems Iris showed slides she had taken of Ali. After a few moments one of the clinicians said he did not think that Ali looked dysmorphic, and several others agreed. Iris responded by asking, "Yes, but is he dysmorphic for a Turkish person?" Her remark points to the Dutchness of normality and suggests that clinicians operationalize understandings of normal Dutch, abnormal Dutch, and a category of everybody else in the diagnostic process.

Iris's presentation of Ali's case demonstrates that physicians see morphology to some degree as nationally, ethnically, and racially bound. These understandings can become explicit factors in the diagnostic process. In these cases, the definition of normal begins with the visual assessment of the subject's morphology. That assessment is then characterized with reference to distinct national identities. The ability to visually characterize morphology as normal with respect to a unique national type enables clinicians to extend their characterization of the subject as normal or abnormal to the genetic level. In such cases nation, race, or ethnicity, or all three, become further categories for defining ranges of what constitutes normal.

The explicitness of Ali's case sheds light on a case that Rob, a senior clinician, presented in which he introduced national background. Rob began his presentation by explaining that his patient was three months old when he first saw her. He then presented a slide of the girl. As soon as the slide came on the screen, Saskia, taken aback by the girl's appearance in the photograph, let out an audible gasp. She told me later that the child's appearance made her think immediately of a possible diagnosis that she then quickly ruled out. The child looked completely normal to me except that she has enormous eyes. Rob said he did not know how well, or if, this child could see because his examination of her showed that she had very little reaction to visual stimulus. He said he was not sure if this lack of reaction was because she was so young or if it had to do with a problem, that is, an organic, presumably genetic problem. He told us that he had moved his hand right in front of her eyes and taken a picture of her with a flash very close up, but there was no reaction. Bart asked him lightheartedly if he had any other weapons he could have used, and everybody laughed. Rob explained that tests showed there were some imbalances in the urine, including a high level of calcium, and the family pediatrician came up with a diagnosis of Williams syndrome. Rob and the other geneticists do not agree with the pediatrician's diagnosis. One of the clinicians asked whether the child looks like the parents. Rob showed the next slide, a profile of the baby, putting off the question, and the next one, which was a slide of two photographs, one of the baby and one of a somewhat older child who also has very big eyes. Rob told us that the second child is a paternal cousin. He said that according to the parents this cousin does not shut his eyes all the way. Several clinicians wondered if Rob could examine the second child, and he told us that the cousin lives in Corsica and that the father of the patient is Corsican. Saskia asked if Rob knew how things were going with the cousin, and Rob responded that according to the patients' parents the cousin's development is apparently completely normal. He said, "The parents say that since [they think] this cousin is normal they do not think there is anything wrong with their child and that they have nothing to worry about." Rob said he had told them, "Well, I do not know if you have nothing to worry about or not."

Rob thinks that the size of his patient's eyes lies outside any possible variation that would fall within the range of normal. He has not, however, found a genetic syndrome that explains the child's appearance and her apparent visual impairment. He also is unable to make sense of the parents' understanding that the cousin has similar eyes and is developing normally. Although Rob himself does not suggest that the patient's unusual eyes can be explained by her nationality, he does introduce the information to his colleagues at the audit because the clinicians consider it potentially relevant information that might assist them in their efforts to find an appropriate category for this child.

Hans made the connection between morphogenic features and ethnic or national identity explicit in discussing with me the relationships between clinic and laboratory in the process of diagnosis. He explained that as researchers isolate more genes associated with genetic disorders on the molecular level, more diagnoses will be supported by laboratory test results. Nevertheless, Hans still envisions an important role for clinical analyses of morphogenic features. He makes this point clear in the comment quoted earlier in this chapter, in which he constructs ethnicity as both a social and a genetic phenomenon by raising the question of whether children who have parents from different national or ethnic groups ("People from Tunisia have children with people from Ireland") are normal or not. With these comments Hans introduces transnational movement into the social context of genetic practice. His observations reflect the tremendous influx of new ethnic groups into the Netherlands resulting largely from the process of decolonization. Ethnicity here presents a morphogenic problem that, in Hans's view, can be resolved only by the expertise of clinical geneticists. Ethnicity is interesting to Hans because it generates new problems in the form of unusual combinations that clinicians can decode and categorize. At the same time, Hans's comments reveal the way clinical expertise becomes defined and its authority is bolstered, in part through reference to social categories of ethnicity, race, and nation.

The dynamics of these processes of categorization are highly relevant in the practice of medical genetics, in which individuals exhibit anomalies that stigmatize them and set them outside group norms. Taking people that are seen as abnormal, then measuring and photographing them in order to place them on a grid does more than produce a plethora of diagnostic categories. Diagnosis also serves the purpose of making the abnormal ordinary by disaggregating the categories of normal and abnormal from healthy and sick, at the same time that it reproduces Dutch social values about ordinariness. The activities I have described in the audit illuminate the social work that goes into distinguishing between normal and abnormal as well as between Dutch and other through the mobilization of ideas about ethnicity, race, and nation. The problems associated with making such distinctions—such as the large overlap between normal and abnormal in some cases as well as the problem of ethnicity—arise from the ambiguities involved in reading photographic images and inconclusive laboratory results. The normalizing processes at work in the audit are especially potent and salient in the Netherlands, where ordinariness is a social ideal. Under these circumstances a powerful dynamic develops between social and scientific values, one through which bodies that are perceived as both normal and abnormal are produced within categories of ordinariness.

In the context of Dutch genetic practice it is important to distinguish between the concept of ordinariness, having to do with *gewoon*, and the idea of being physically normal, having to do with *normaal*. Whereas the opposite of gewoon is simply *ongewoon* (unusual) or *buitengewoon* (extraordinary), the designation *niet normaal* (not normal) connotes some sort of defect. One informant explained that she might describe someone who was missing a limb or who was sick as niet normaal. She also suggested that the opposite of normaal is *gek* (crazy/strange), a concept carrying much stigma. In the normal course of events a person with a genetic disorder would be considered niet normaal and therefore unable to be gewoon. Once situated within the category of a known syndrome by Dutch genetic practice, however, such a person becomes normaal for that syndrome. Her difference ceases to be random, and her status ceases to be solitary. As part of a bounded group of people with shared characteristics she becomes capable of assimilation into a social category in which she is ordinary. This process echoes the social tradition of pillarization, which manages difference by bounding it in recognized social categories and containing it by promoting conformity within groups.

Although the presentation of patients at the audit serves the purpose of producing diagnoses, it also facilitates the more subtle goal of bounding difference and enabling the achievement of the Dutch social ideal of ordinariness. The emphasis people coming through the clinic place on fitting in illustrates this ideal. For example, in discussing laypeople's reactions to clinical diagnoses with me, Iris reported a case that demonstrates the importance of ordinariness. In this case, she informed a couple that the laboratory had diagnosed a sex chromosome condition from the woman's amniocentesis. She told them that the fetus had an extra Y chromosome. On the basis of studies of other children with this genetic anomaly, Iris explained what they should expect if they chose to continue the pregnancy and have the child: Their child would look like a normal male, but he might have learning difficulties and would not be very intelligent. Thinking that she was delivering bad news, she was surprised when they responded, "Oh, that's all right. We don't mind if he isn't very intelligent. We aren't very intelligent either. It would be very difficult for us to have a child who was very clever. It would be worse for us if you told us that we should expect a child that is very smart." This last case reveals the significance of the developmentally disabled people and the extraordinarily brilliant granddaughter in the film *Antonia's Line*. In chapter 1 I argued that the film portrayed Antonia as embodying the Dutch social ideal of tolerance by surrounding her with people who were different. Her understanding and acceptance of those differences were what made her an idealized symbol of modern Dutch identity. Iris's story shows that even high intelligence, a difference that many would consider to be positive, may be interpreted as a problem where it signals difference. It further challenges assumptions that, given the choice, everyone would want their children to be very clever.

A number of parents of children with genetic conditions who spoke with me mentioned their children's ordinariness within a particular category. One mother explained that when she sees pictures of children with Williams-Beuren syndrome, the same developmental condition her daughter is diagnosed with, she recognizes her daughter in the appearance of the other children. She went on to say that children with this syndrome "all look like each other—they could be from the same family." Another parent of a child with Williams-Beuren syndrome said, "They look like they all have the same father." These parents, having children who did not fit the ideal of normal, stress that their children become normal, perhaps even kin, within another category. Their comments normalized the children not just by placing them in a diagnostic category, but also by making that category into a community. These examples of parents' talk about and reactions to genetic conditions in their children disclose the dynamic production of diagnostic categories and the contested meanings of such categories. For the clinicians, the categories of genetic anomalies represent pathology. Parents, however, take categories that are presented to them by experts and imbue them with their own, often surprising or unexpected significance.

Rayna Rapp has described how American genetic counselors, pregnant women, and parents of children with Down syndrome normalize those children by speaking of similarities among them, similarities through which they construct kinship relations.[14] Rapp gives examples of how understandings of alien kinship among people dealing with Down syndrome can romanticize difference, can be used to construct newly imagined families, and may also blur boundaries between disabled children and other species. My material suggests that in the Dutch context parents do this as well, but in stressing similarities among people with shared syndromes they are not just imagining kinship but also are producing a socially valued ideal of normal from what is perceived as abnormal. Part of what enables them to successfully see ordinariness in their children is the diagnostic categories produced through the work of medical geneticists.

The explicit, complex, and contested dynamic of normalization at work in the audit produces multiple categories of ordinary that exist simultaneously. Rather than simply classifying an individual with a genetic anomaly as abnormal, clinicians strive to fit the person into a scientifically or medically defined category in which they may be perceived as normal and even healthy, thereby reducing the social threat of their conditions and facilitating their ability to partake in the Dutch social ideal of ordinariness.

The communication between medical geneticists and general practitioners highlights this process. In a letter explaining his diagnosis of Klippel-Feil syndrome to Ineke's pediatrician, Bart lists her dysmorphic features at length, telling the pediatrician that the relationship between Ineke's height and span are disproportionate (though because of the neck and shoulder abnormalities the measurements may be unreliable), that her right ear is a bit smaller than the left, her palate is normal though there is some crowding of her teeth, she has narrow shoulders, she favors her left arm, there is a light clinodactyly but the mother also has this in a milder form, and she has a small head circumference (microcephaly) that does not correspond with the head circumference measurements of her parents. Nonetheless, Bart begins the letter by saying that "Ineke is not a sick girl." The practice of making what appears abnormal ordinary, as I observed it at the genetics center, does more than produce a plethora of categories in which to place people. Bart's letter to Ineke's pediatrician explicitly decouples the categories of normal and abnormal from those of healthy and sick. By explaining Ineke's symptoms and emphasizing that she is not sick, Bart places her in a category in which she is normal in relation to others within that category. This bounding does not deny genetic anomalies but layers them with a secondary categorization known as a syndrome within which syndrome-bearers become ordinary.

PRODUCING THE ORDINARY

In their study of Foucault's work, Hubert Dreyfus and Paul Rabinow elaborate upon the notion of normalization and its relation to science. They write that for Foucault "norms are always on the move as if their goal was to bring every aspect of our practices together into a coherent whole. . . . [N]orms do not rest but, at least in principle, are endlessly ramified down to the finest details of the micropractices, so no action that counts as important and real falls outside the grid of normality. In addition, as in normal science, the normalizing practices of bio-power define the normal in advance and then proceed to isolate and deal with anomalies given that definition" (Dreyfus and Rabinow 1983:258).

The dynamic processes of diagnosis I have analyzed in this chapter make it clear that, in the practice of genetics in the Netherlands, norms are never at rest. More than this, however, beyond the conception of an articulated norm evolving in relation to categories of abnormal, my examination of Dutch experiences with genetics reveals how the desire for ordinariness can lead clinicians to articulate multiple conceptions of normal, each dynamic in its own right, each coproduced in relation to the others as diverse normalities become a function of unfolding diagnoses and categorizations of humanness.

Dutch clinicians use the implicit, unarticulated, and, therefore, unmarked status of normal to identify and categorize genes and the individuals bearing those genes. The processes at work in the audit reveal the dynamic interplay among medical practitioners' desire to identify and pathologize difference, the Dutch emphasis on tolerance, and Dutch values about ordinariness. This interplay takes place through the interpretation of bodies perceived as genetically abnormal. Abnormal bodies threaten social disruption specifically because they stick out—their genetic anomalies make them visible. Without medical interpretation there is no way to perceive such deviant bodies as other than simply abnormal because there is no popular category of normal into which they can be placed. The audit and the clinicians' interpretations of genetic disorders do not deny the category of abnormal but layer it with a secondary categorization known as a syndrome. Within these syndromes, syndrome-bearers are normal in relation to each other; and their appearance and behavior are normal for that category. Once such a category is bound and defined, Dutch values centering on tolerance require that those within the category be considered ordinary. In facilitating the attribution of ordinariness, medical interpretations help to assimilate otherwise threatening anomalies into a common Dutch ideal of ordinariness, thereby minimizing the threat of such anomalies and reshaping social understandings of genetic disorders. This process gives rise to an alternative to simply opposing genetic anomalies to a pregiven definition of normal through allowing a secondary categorization of ordinary within a specific, known category.

The audit demonstrates how local cultural values surrounding ordinariness interact with medical and scientific imperatives to name, categorize, pathologize, and normalize difference. Laypeople, such as parents, may also contest and reshape the meaning of genetic difference as they make sense of genetic conditions. Finally, the interaction between local Dutch values, science, and medical practices in the Netherlands illustrates the need to break down the monolithic treatment of Western science. As Dutch culture informs genetic knowledge and practice, it distinguishes them from the genetic knowledge and practices of other societies that, together with the Netherlands, are usually glossed as Western.

Chapter Four

BACKWARD AND BEAUTIFUL

CALVINISM, CHROMOSOMES, AND THE PRODUCTION OF GENETIC KNOWLEDGE

It is between 8:30 and 9 a.m., and Bart, a junior clinician, is raising a question with his colleagues at the weekly audit. Bart explains that he has a couple (whom I will call Veronica and Joop) whose daughter has cystic fibrosis and that they had also had a second daughter with cystic fibrosis who recently died. The consensus among the local physicians with whom the family dealt was that the deceased child died because of complications from cystic fibrosis, but Bart does not think cystic fibrosis explains this child's death. In fact, given the records on the child and her death, Bart suspects it may have been caused by a second recessive disorder. He then tells us that the couple "say they are not consanguineous, and they became very angry when I brought up the possibility. They are from Zeedrecht, but since they have two different mutations it is not likely that they are consanguineous and they say they aren't." The couple, he explains, wants to have another child and wishes to ensure they do not have another child with cystic fibrosis, yet they do not want to terminate a pregnancy and therefore do not want to have amniocentesis for prenatal diagnosis. As with most questions posed in this part of the meeting, the issue is less about how to treat the patient(s) medically than about how to deal with a particular case procedurally.

Veronica and Joop contacted the clinic seeking a specific solution: they wanted a preimplantation diagnosis for cystic fibrosis. An experimental program in preimplantation diagnosis had recently begun in Maastricht at one of the eight Dutch genetics centers. This procedure involves taking one cell from eight cell embryos created through in vitro fertilization, screening them for a particular genetic disorder, and then implanting only the embryos found to be free of the anomaly in question.[1] Having only one cell to work with in producing a diagnosis makes preimplantation diagnosis quite intricate; since it is still in an experimental stage, those involved in the project have strict criteria for inclusion in the program. The criteria for this clinical trial, as in any well-designed clinical trial, were established to prove whether or not preimplantation diagnosis would actually work. At the time of this trial the human genome had yet to be completely mapped. Not only were many genes not understood, but many had not been found or delineated. In order to prove that preimplantation diagnosis worked, researchers included people in the trial only when they could be certain they could find the gene they were concerned about. This meant the relationship between genetic mutation and its manifestation in a genetic condition needed to be well understood, and the gene itself had to be delineated well enough that the researchers would know where among the three billion base pairs of the genome they would find it. Enrolling people in the trial whose cases did not conform well to the trial's criteria could compromise the potential success rate of the trial itself and thus undermine the ability of the researchers to prove the efficacy of the procedure. In the end, the procedural question Bart was posing to his colleagues was whether or not they could tell the researchers running the preimplantation trial precisely what to look for. At the time, doing preimplantation diagnosis even for cystic fibrosis—one of the most commonly occurring genetic conditions—was quite complicated. As Bart points out, "Preimplantation for cystic fibrosis is an experimental technique that has been done very few times and never before in Holland."[2] But, as Bart continues to present his case, it becomes clear that he thinks Veronica's and Joop's case is even more complicated.

Bart explains that the couple very much want to have preimplantation diagnosis, but he is not sure whether it can be successful in this case because of his questions surrounding the causes of their child's death. The problem for Bart and for this couple is that if there is, as Bart suspects, a second, unknown recessive disorder, then this couple is not a good candidate for preimplantation diagnosis because the clinicians and researchers could not know what to test for. Bart tells us that the couple could go to Belgium, where preimplantation diagnosis is available on a commercial basis.[3] The problem with this option is that going to Belgium would cost about eighteen thousand dollars, an amount the couple would likely have to pay themselves. If they qualify for the experimental program in the Netherlands, the process would be covered by the complicated combination of private and public health insurances that cover all legal residents in the Netherlands. Bart tells us, "We have to decide what we think about this case because it is very complicated. Do we proceed? Do we advise the parents to proceed? Should we write to the insurance company and ask them to pay and maybe they will [pay], or meet them halfway? . . . The patients have to be referred, and we have to be honest with Maastricht or Brussels." One of the senior clinicians again raises the question of consanguinity, and Bart reiterates one of his original points, saying, "They say they are not and they became very angry when I talked about it, but they are from Zeedrecht. Her father is Jewish though and [originally] from Amsterdam, so maybe they are right."

As I listen to Bart explain this case, I puzzle over the repeated references to Zeedrecht, religion, and consanguinity, wondering what he is communicating and what the clinicians are comprehending in his comments. Like everyone else in the room I realize we are talking about an extraordinarily unlikely scenario. We all are well aware of the clinical presumption the clinic director articulates as "one child, one problem, most of the time." What is the meaning of the statements being made in this discussion? Why are these statements being made and not others? Why is Bart even considering the possibility of a second recessive disorder? What is it about this couple that makes it possible to think this? Why is Bart questioning the patients' understanding that they are not consanguineous? What is it that is making them angry? What ideas are the clinicians bringing together in their attempt to make something that makes sense to them as truth? Exploring these questions requires untangling the ways the clinicians are mobilizing intersecting ideas about purity, consanguinity, modernity, genetics, science, identity, geography, and religion in the way they are evaluating Veronica's and Joop's case. In the process the entanglement of biological and social facts comes into sharper focus.

In this chapter I argue that knowledge about social aspects of local geography, biology, and religion becomes mutually implicated with scientific knowledge in producing clinical interpretations of the body. I focus principally on the population of Gereformeerden, or Orthodox Reformed Calvinists, a conservative religious group who are associated with a bounded and well-defined region in the Netherlands. A shared understanding of this group as backward can be and is used as one measure against which other Dutch people may assert a sense of their own modernity. Together with clinical and laboratory evidence, medical geneticists in the Netherlands will call upon common Dutch understandings of geography and social practices associated with religion—what one might call a geographical imaginary—to support suspected diagnoses. I show that in such practices, these clinicians produce not just a geographical imaginary but also a biological imaginary.[4] In developing such imaginaries, the clinicians invoke a distinctive aesthetics in which the genes produced in spaces and bodies imagined as socially backward are interpreted as beautiful insofar as they present the clinicians with interesting, unusual data that can be developed into aesthetically compelling diagnoses. These processes critically shape the development, understanding, and uptake of knowledge about genetics in the Netherlands. I further elaborate the ways Dutch genetic practices are distinctively informed by Dutch social life by showing the processes through which locally shared understandings of social life condition both clinical practices and the content of medical and scientific knowledge. In turn, I also consider how this dynamic refracts back upon Dutch society, affecting understanding of Dutch identity and social life.

Exploration of the issues raised by the presentation of Veronica's and Joop's case throws into high relief how biological facts have no meaning in and of themselves but instead become meaningful only in relation to social facts. This chapter contextualizes Veronica's and Joop's clinical encounter by elucidating the circulation of social and biological facts both within and beyond the clinic. The illustration of such processes in action reveals that social facts can become biological facts and that biological facts can become social facts, thereby sharpening one's understanding of how science and society are coproduced through human activity. Donna Haraway's introduction of the word-concept "naturalcultural" reminds one that the categories of nature and culture are often inadequate for describing technoscience and its myriad effects in the world.

THE NATURALCULTURAL PRODUCTION OF BIOLOGICAL AND SOCIAL FACTS

When I talked to Bart after the meeting in which he presented Veronica's and Joop's case he explained that Zeedrecht is one of the villages in the Netherlands known for its strict religious practices. For the clinicians, this social fact immediately reinforced any suspicions of consanguinity. While we were talking Bart guided me toward the map of the Netherlands hanging in one of the hallways of the clinic; he showed me where Zeedrecht was located, explaining that there were a number of religious villages in that part of the country. Pointing at the area, he said, "There is a lot of consanguinity in this area."

In talking about Zeedrecht, religion, and consanguinity, Bart is talking about the Gereformeerden sects in the Netherlands. These churches are distinct from the mainstream Calvinist church to which much of the Dutch elite has historically belonged. In spite of the rapid secularization the Netherlands has experienced since the 1960s (Ellemers 1981; Middendorp 1991), a number of Dutch villages widely known for having populations that promote and maintain strict religious observance still exist. These villages compose what is popularly known as a Bijbel Gordel (Bible Belt). Many, but not all, of those who live in these villages belong to Gereformeerd churches. They are popularly called Zwarte Kousen (Black Stockings) because of the black hosiery they wear. By invoking Zeedrecht, therefore, Bart is asking his audience to imagine an array of attributes associated with this geographical region of the Netherlands.

Members of the various Orthodox Reformed churches have never exceeded much more than 10 percent of the Dutch population, and currently make up only about 8 percent of it (Shetter 1987:169–70). Nevertheless, they serve a number of important symbolic roles in contemporary Dutch society. Since they made up the vast majority of the Dutch resistance during the Second World War, there is a level at which they are honored national heroes—a status which they leveraged into significant political power in the postwar period. At the same time, their refusal to compromise during the Nazi occupation necessarily confronts other Dutch people with the fact that so many others did accommodate German occupiers. Furthermore, although the Gereformeerden have always maintained a distinct and minority religious orientation, there is a sense in which many Dutch people see in the Gereformeerd orientation to religious life something of what life was like for all Dutch people until quite recently. In this sense the Gereformeerden represent a past in which social life for all Dutch people was oriented around religion. For most Dutch people this is not the distant past of the seventeenth-century Golden Age, when the Netherlands was a world power, but a more recent past of conservatism and constraint that extended well into the twentieth century, as depicted in the film *Antonia's Line*. In these ways many Dutch people perceive the Gereformeerden as other in both time and space. Such understandings of the Gereformeerden serve as a symbolic contrast against which mainstream Dutch people conceptualize themselves as modern. This role is demonstrated in the way people both inside and outside the clinic talk about the people who live in these towns, presenting their beliefs and practices as backward at the same time they distance themselves from those beliefs.[5]

Take, for example, the comments of Will, a geologist and a lapsed Catholic. One afternoon he told me that "in the Netherlands the Catholics are the liberals. It is the Protestants who are really conservative." Will emphasized that the conservatism in some of these villages is rather counterintuitive:

> **Will:** In some cases the people in a village require premarital sex.
> **KST:** Really! Why is that?
> **Will:** Because in some villages a woman cannot be married until she proves that she can have children.[6]

Will went on to tell me about the protests against the pope during the papal visit to the Netherlands in 1985 (see Crump 1985) and about the religious villages that make up the Bible Belt. Will linked certain behaviors to religious values, explaining that in these villages people dress in traditional clothing and are not allowed to ride bikes on Sundays. When describing these same religious villages to me, thirty-year-old Rudie said, "You cannot drive down the streets in those towns on Sundays because the people walk in the middle of the street, and they do not move out of the way because they think you should not be driving on Sunday." When I asked whether or not they think Dutch people have common qualities or characteristics, sixteen-year-old Annemiek and her eighteen-year-old classmate Marc spoke at great length about Dutch tolerance and openness. At one point, discussing the existence and consequences of these ideals, Annemiek set apart people in the religious villages: "Except the Black Stockings Church [Zwarte Kousen Kerk]. They have, of course, a unique, separate belief in the Netherlands."

The above comments demonstrate that many Dutch people perceive places associated with religious devotion as homogeneous as well as socially unique and geographically remote. These villages are perceived as being marginal not only by outsiders, but also by at least some of their residents. Astrid, a fifty-year-old physician who was raised in a religious family in a small village not far from one of Holland's largest cities, described her childhood experiences there:

> I was born in a village, a small village, below the big rivers, as we say . . . and it was a very small village with a lot of water around it. So we were a little bit isolated. And when I was a teenager they built a big bridge. Then it was called a big bridge, now it is [considered] a small one. Before that time, if you had to see a [medical] specialist then we went by boat. . . . I went to school, and it was not normal for a girl to go to high school. Especially if your parents had not been to high school. . . . After elementary school I went to the high school in Waalenchem, and in the town it was a scandal because I was a girl. . . . I went to the high school [and] afterward other children could go also. . . . I was the first.

Astrid's memories of her experience growing up, like the more distanced contemporary descriptions people offered of religious villages, suggest a temporal disjunction from mainstream Dutch society.

In his work on marginal places, the geographer Rob Shields stresses that

> marginal places, those towns and regions which have been "left behind" in the modern race for progress, evoke both nostalgia and fascination. Their marginal status may come from out-of-the-way activities, or being the Other pole to a great cultural centre. . . . [G]eographic marginality . . . is a mark of being a social periphery. That is, the marginal places that are of interest are not necessarily on geographical peripheries but, first and foremost, they have been placed on the periphery of cultural systems of space in which places are ranked relative to each other. They all carry the image, and stigma, of their marginality which becomes indistinguishable from any basic empirical identity they might once have had. (Shields 1991:3)

Many of the smaller villages associated with religious devotion, like the one where Astrid grew up, are within ten to twenty miles of major cities. Today at least they seem less geographically remote than socially marginal. Nevertheless, their marginality has significant social power.

Anthropologists have long shown that time and space are socially produced phenomena having multiple consequences for the organization of distinctive societies.[7] Less attention has been paid to the dynamic role played by deeply embedded understandings of geographically specific social practices, real or imagined, in the daily interactions that simultaneously produce people and their worlds. Such geographical imaginaries emphasize that understandings of behaviors that are geographically located are a product of the way people imagine the world, whether or not what is imagined actually occurs in social practice. These imaginaries can have enormous social power, as we see in Bart's mobilization of a geographical imaginary and an associated biological imaginary in his presentation of Veronica's and Joop's case.

Some of the villages making up the Bible Belt are quite small and relatively homogeneous while others are larger and more diverse. Amersfoort, which one physician described to me as part of the Bible Belt, is a medium-sized city with a diverse population. Its size keeps Amersfoort from being popularly categorized as particularly religious in the geographical imaginary, but its proximity to villages that are widely known for their religious devotion still can make its location clinically significant.

The village of Bergwijk, where I conducted fieldwork in a high school, is quite small, and a number of people told me it was known for its religious population. Walking down the streets of the village or exploring the weekly farmers' market, I often saw women wearing the head scarves typically associated with Moroccan immigrants and their descendants doing their daily shopping. There are state-supported religious schools in Bergwijk; I interviewed students at the nondenominational school. While some of the students at the school may very likely be devout, the students I interviewed never conveyed any religious orientation either in their dress or in their comments, and they appeared to be quite secular. The kind of diversity found in Amersfoort and Bergwijk is often elided in the way that people talked about the Bible Belt villages and also, as I explore below, in the clinical presentation of people from places known for religious devotion.

RELIGION, GEOGRAPHY, AND GENETICS

People both inside and outside the clinic described a constellation of values surrounding punishment, shame, and privacy regarding genetic conditions. Many informants explicitly linked these values to traditional religious beliefs, reflecting themes central to Max Weber's classic study of Protestantism and capitalism. Weber's thesis linking the rise of capitalism to Calvinism is contested, particularly in the Dutch case.[8] Nevertheless, his elaboration of the disquiet caused by the Calvinist doctrine of predestination is significant here. Weber emphasizes that the concept of predestination—that one is saved or damned before birth and that no worldly activities can change the fate that awaits one after death—is a defining feature of Calvinism. He argues that the idea of predestination continuously raises the question of whether or not one is a member of the elect, whether one has grace, and a concomitant desire to establish means of knowing one's destiny (Weber 1958 [1904]:110). According to Weber, success, attributed to leading an exemplary life, became a sign of one's grace not only for oneself but also for one's community (1958 [1904]:114–17).

Anthropologists have long recognized that people everywhere have culturally appropriate means of dealing with what are perceived to be anomalous or even monstrous births (Douglas 1966; Evans-Pritchard 1956; Morgan 1989). Although as individuals they distanced themselves from it, the Dutch people I met repeatedly introduced the idea that for many in the Netherlands the anomalous births many now recognize as involving genetic conditions used to be, and in some cases still were, interpreted as punishment from God and thus as a sign of sin or of one's lack of grace or both. This interpretation of genetic anomalies combined with other social values about fitting in or being ordinary—although less significant to the mainstream secular population—worked to impose a silence around genetic conditions in the Netherlands. As the comments below illustrate, historically people both hid and refused to speak about the presence of people with genetic anomalies. Such silence about genetic anomalies remains powerfully salient for many in the Gereformeerd population, many of whom continue to interpret genetic anomalies as signs from God. This silence affects the practice of medical genetics because the Gereformeerden are the very people whose bodies are perceived by geneticists as having the most to tell or to reveal about genetics. In this belief the functioning of the biological imaginary is most manifest. That is, in identifying a person as being from the Bible Belt, geneticists are imagining an array of historical conditions likely to be associated with a small, consanguineous community deemed to be geographically and temporally distant from modern Dutch society.

When I asked people about the role of religion in the context of genetics, they frequently brought up the Veluwe, an area—part of the Bible Belt—known for its religious villages. Regina, a thirty-three-year-old office worker, and her husband, Paul, a carpenter, had a lot to say about the relationship between religious beliefs and genetics:

KST: Do you think religious beliefs influence how people in the Netherlands think about genetics and biological inheritance?

Regina: Yes, I think so. I am glad that I was raised Catholic. On the other side [in the Veluwe] with things like genetic counseling, then you stick your head in the sand and you say that it is God's will that you got/had an unlucky child or something. . . . They say it is God's will . . . it is like with polio.

Paul: Yes, like what happens with polio. They won't get vaccinated.

Regina: The parents choose for the children, and the children cannot choose for themselves.

KST: When did this happen with polio?

Regina: A couple of years ago there was an epidemic in the Veluwe. . . . They said, "We do not want to do that [have vaccinations]" . . . we read about it in the newspaper. That [the resistance to vaccination] is really typical. . . . I think that it is very bad.

This couple distanced themselves from the Gereformeerd population while speaking of them in relation to geography and modernity. At the same time, they introduce the idea of disease as a manifestation of God's will.

The theme of genetic anomalies as punishment from God and something to be ashamed of came up repeatedly in my interviews. I asked Carla and Gerard, a couple in their late thirties who have two sons, one of whom has Down syndrome, about their understandings of the source of genetic conditions. I asked, "What kinds of explanations have you heard for inherited illnesses or abnormalities, for how they begin?" They responded with a variety of explanations: "We heard that it was a punishment [*straf*]" and also that genetic anomalies could be the result of social misbehavior, such as sexual promiscuity or drug use by women during pregnancy. They then shifted to talking about what they described as the scientific explanation and talked about chromosomes and biology lessons from school. When I asked them if they think religious beliefs influence how Dutch people make sense of genetic inheritance, they told me that "those who are strong believers do not believe in evolution . . . they see genetic abnormalities and illnesses as punishment from God." Like Regina and Paul, Carla and Gerard mentioned the polio outbreak of a few years earlier and the refusal of the Gereformeerd population to be vaccinated. They said that "strong believers do not have the nicest reactions to genetic abnormalities" and described the case of a minister who told the parents of a child with a genetic condition that the disorder was a punishment from God: "That still happens. It also still happens that there will be a disabled or mentally retarded child on a farm that nobody knows about because it [a genetic condition] is considered shameful." This last comment might be interpreted simply as an urban perception that rural social life produces a culture different from that found in the metropole. In this context, however, Carla's and Gerard's comments also reflect the geographical imaginary operating in people's understandings of religious devotion. The Gereformeerden are (or at least are imagined to be) concentrated in the small, rural villages making up the Bible Belt. Hence, the last comment, which serves to distance the speaker from communities perceived as backward and premodern, illustrates the conflation of religion, geography, and science.

Kitty, a sixty-three-year-old woman with an autistic grandchild, also connected religious beliefs to attitudes about genetic conditions:

> **KST:** Do you think that religious beliefs influence the way that people in the Netherlands think about genetics and biological inheritance? . . .
> **Kitty:** Now, it often happens that people who go to church often abhor [*verafschuwen*] or shun [*schuwen*] an unlucky family with an unlucky child.[9]

Later in the interview Kitty returned to her ideas about religion and genetics when discussing her knowledge of genetics:

> **KST:** Have you read or heard about genetics or biological inheritance in the newspapers or on TV.?
> **Kitty:** Yes.
> **KST:** Do you think that knowledge about genetics and biological inheritance influences the way you think about your body or what happens in your body?
> **Kitty:** No. I do not think so. No . . . I think that there is also too little read because there is still very little written about it. It is a taboo subject . . . and it [information about genetics] is not publicized well enough.
> **KST:** And why do you think that it is a taboo subject?
> **Kitty:** I think that it earlier had to do with religion. . . . Not everyone is ignorant . . . [but] . . . people do not want to talk about it . . . you have a particular illness and people do not want to talk about it. I cannot mention it . . . it is dead-on normal [*dood normaal*]—everyone can get [sick] . . . and you should also be able to just talk about what to do . . .

> [but] in the Netherlands . . . people would rather not talk about it. Very often people are not happy to know what [illness] another has.

Kitty went on to speak at length about how genetic problems are perceived as being shameful and something one could not talk about, but she also emphasized her disapproval of this social norm. She stated that genetic anomalies were neither punishment from God nor something for which people should feel responsible. She argued that these norms should be changed and that people should talk more freely about their experiences with illness and disease. For Kitty and for a number of other people I interviewed religious explanations for illness, disease, and embodied difference are no longer acceptable. Such explanations were widespread in the Netherlands until very recently, and vestiges of the attitudes such explanations provoked can still be found. Those, like Kitty, who have a person in their family with a specific embodied condition are acutely aware of the persistence of these attitudes in the Netherlands today. Rejection of such attitudes, and the religious explanations on which they rest, in favor of medical and scientific explanations is part of enacting and articulating a modern Dutch identity.

The above examples illustrate two related social processes. The first is a geographical imaginary, the spatial sense Dutch people have about where religious devotion is a significant component of social life. Embedded in this imaginary is an elaboration of particular ideas about what that religious devotion entails, including a preference for endogamy, a resistance to certain forms of medical intervention, and a persistence of understandings of embodied conditions as being signs from God. In this elaboration one sees the emergence of a biological imaginary that, as I discuss below, is highly significant in the context of genetic practices. The second is the contrastive symbolic role that people within these religious spaces serve in facilitating the ability of mainstream Dutch people, including Dutch geneticists, to produce a sense of themselves as modern subjects. In such processes one can glimpse a specific instantiation of the dynamic Peter Stephenson describes in articulating Dutch concepts of the self as compressing "two contradictory polarities—I am just like everyone else/I am unique" (Stephenson 1989:232).

In the Netherlands the importance of fitting in by being ordinary historically combined with Calvinist values to produce a resounding silence around genetic conditions. Such conditions marked individuals as existing outside the realm of ordinariness. At the same time, genetic conditions were widely perceived as being punishments from God and therefore as something for a family to hide out of shame. Scientific explanations thus may work to replace religious explanations of genetic conditions. In offering such explanations, scientists articulate categories in which to place those with such conditions. In bounding genetic difference through these categories, medical and scientific practices render genetic conditions less threatening. As a result, the silence around these differences has begun to lift. Rejecting religious explanations for genetic anomalies in favor of medical and scientific explanations is perceived as a sign of modern Dutch identity because people in the Netherlands often associate religion with lack of sophistication or with backwardness and because, as I show below, religious backwardness is defined in part in terms of genetics.

GENETICISTS INTERPRET RELIGIOUS PRACTICE

Medical geneticists operationalized the geographical and biological imaginaries in their daily practices in powerful, nuanced ways. Rob Shields's argument that "social divisions are spatialised as geographic divisions and . . . places become 'labelled,' much like deviant individuals" points to the significance of this imaginary. Shields explains that

> Habits such as spatialising important conceptual oppositions (for example, putting one thing on the right, and another thing on the left, or classifying people by the places they come from: the "right" or the "wrong side of the tracks") have been studied as pathologically irrational forms of behaviour but [are] . . . an essential conceptual shorthand. These prejudices amount to a form of everyday knowledge which has been trivialised and dismissed by researchers interested in more "serious" knowledge. Nonetheless it betrays a systematic "disposition" toward the world (cf. Foucault 1980a; 1980b; 1982) coded in the framework of common sense. (Shields 1991:11)

In interviews and conversations medical geneticists regularly employed a Dutch geographical imaginary as a conceptual shorthand in their daily practices. As people employed this shorthand in thinking and talking about genetics they produced not just a geographical imaginary but also a biological imaginary. That is, as I elaborate below, they projected onto an imagined geographic space specific social practices that, in their minds, had specific biological implications and results.

Diana, a senior medical geneticist, raised several of the central beliefs associated with religion in the Netherlands when I asked if she thought there was anything unique about the practice of genetics in the Netherlands. In responding to my question, she drew comparisons between Gereformeerden Dutch people and Moroccan, Turkish, and other ethnic and religious minorities residing in the Netherlands. Her comparison demonstrates how secular Dutch people can perceive devout Dutch Protestants as distant or other. It also points to connections among genetics, religion, and national identity. She began to answer my question about genetic practices in the Netherlands by talking about restrictions on embryo research and then shifted to talking about religion:

> **Diana:** I do not know if you know that Dutch people are really conservative and religious—some groups—and they do not want to have this [genetic counseling]. They do not like to have the possibility to decide before God because if you decide . . . then you decide for God. They think it is the right of God to decide if a child is to be born or not . . . you may not interfere in God's will. That is the opposition.
>
> **KST:** So the opposition mostly comes from religious groups?
>
> **Diana:** Yes. And maybe you have also seen that there are many foreigners in Amsterdam. They come from Turkey, Morocco, China, Surinam, they are ignorant and they are often very religious and they are not allowed, from what I know, to use amniocentesis . . . [or] . . . to decide for abortion. Abortion is a problem in those cultures and they are very strict. So that is the problem, because the older people do not come here [to the clinic] but their children who are born here . . . they want it . . . and we see this often in Dutch people with the older generation. They have a problem if their children decide to use amniocentesis or ultrasound to decide whether or not to have a child.
>
> **KST:** So you see it as both a religious issue and also a generational issue and that younger people are more accepting?
>
> **Diana:** They are more modern, they are more used to planning, to talking about something . . . the older ones, they keep their mouths shut. If they have had a miscarriage they do not talk about it. If they have a dead child, they do not talk about it. They don't even mention it. If you ask them how many children they have had they will only count the ones that are still alive . . . so I have also to ask, "Do you have children who have died? Have you had a miscarriage?" and it is still very difficult to

> ask these questions because sometimes people think, "If I had a miscarriage it is a sin; I have been punished by God; I have done something wrong." Sometimes they look for those kinds of reasons so that is why I think they don't like to talk about it—because you might think badly about the person . . . because we work with genetics we have to be informed about the generations. . . . We ask about grandmother, grandfather, and then we hear about things like [when] they say, "My mother has something but I don't know about it because she does not want to talk about it, and she also doesn't want me to be interested in it, [or] to ask her if somebody has something in the family [and] may I investigate it? She doesn't want people to know."

Diana appeared to have a dual purpose in highlighting the silence about inherited disorders and the resistance to new reproductive technologies among the more devout population. First, she explains the social uniqueness of clinical genetics practice in the Netherlands. Second, she marks both Dutch and non-Dutch religious minorities as traditional, unmodern, backward, and other relative to the mainstream and largely secular Dutch population. These religious minorities include both the largely rural Dutch Gereformeerden and the predominantly urban Muslims, primarily of Turkish and Moroccan descent. In comparing the minorities to the mainstream, she is implicitly constructing a Dutch national identity in and through understandings of gene, religion, and geography.

One symbol of these groups' backwardness that is made both explicit and significant in the context of the genetics centers is an elevated incidence of consanguineous marriages. Here the geographic and biological imaginaries are most explicitly at work in the context of genetics. Indeed, the Zeedrecht couple's anger at Bart's questions about whether they might be related to each other can be understood in the context of popular attitudes toward consanguineous marriage. The couple became angry because they know that the social practice of endogamy, with its increased possibility of consanguineous marriage and attendant subtext of inbreeding, is widely perceived as backward. At the same time, the social fact of consanguinity is also widely understood in the scientific community as a resource for expanding genetic knowledge.

The topics of religion and consanguinity came up repeatedly during my fieldwork. One day, for example, while I was driving with Hans and Saskia, two of the clinic's medical geneticists, to an outpatient clinic at a regional hospital in Meerwijk, one of the towns in the Bible Belt, Hans told me that the clinicians get lots of "beautiful diagnoses" (*mooie diagnoses*) from the hospital in Meerwijk. I asked him what a beautiful diagnosis was. Both he and Saskia immediately responded that there was a lot of inbreeding in this area because of the religious values, and because of this consanguinity they "get diagnoses and not guesses or maybes, and that is very nice." He told me, "We get beautiful chromosomes from this hospital because it is in a very religious area. That is why we keep coming here. My most famous patient came from this hospital."

Given geneticists' understandings of simple Mendelian inheritance, in which genetic characteristics are inherited in recessive, dominant, and sex-linked patterns, consanguineous marriages are scientifically and clinically significant. Their significance arises from the increased possibility that recessive traits, including those associated with disease or anomaly, may come forward in the children of those marriages. Geneticists expect an increase in recessive conditions with a consanguineous couple because they have a greater chance of having homologous chromosomes—identical because they originated in a common ancestor. There is a calculus at work here in which endogamy leads to both purity and danger (Douglas 1966): it produces a homogenous society uncontaminated by outsiders along with the lethal potential of contaminating genetic anomalies.[10]

Hans emphasized the medical and scientific value of chromosomes from these communities both in his comments on the way to Meerwijk and in those he made as we left the hospital there. During our visit to the hospital, the staff were somewhat disorganized. Upon leaving, Hans explained that despite the confusion, clinicians from the genetics center continue to make the effort to hold an outpatient clinic because the chromosomes that come from Meerwijk are so valuable. He said, "It is always like that in Meerwijk—they know we are coming but when we arrive they are not ready. But they send us lots of chromosomes so we keep coming." Hans's comments demonstrate geneticists' sense that people from religious villages are resources for scientific and medical knowledge. Regional hospitals and their staff physicians provide medical services to religious villages. Clinic physicians' relationships with physicians at the regional hospitals facilitate the geneticists' access to people in these communities and to their highly desired chromosomes.

BEAUTIFUL CHROMOSOMES

In an interview several months after my visit to the regional hospital in Meerwijk I reminded Hans about his comments about beautiful chromosomes and asked him,

> **KST:** What did you mean by that?
> **Hans:** Yes, lots of beautiful abnormalities. Professionally beautiful, of course, that is not beautiful and you have to realize that. You have to realize that what doctors think is a fun [*leuk*] syndrome, that is fun for doctors, but not for patients. And perhaps that was an area of consanguinity. Perhaps it was Zeedrecht or Meerwijk?
> **KST:** Yes, Meerwijk.
> **Hans:** Meerwijk, yes, there is lots of consanguinity there . . . due to the religious background, I think.

Bart, the clinician who presented the case of the couple from Zeedrecht, also described connections between genetic anomalies, religion, and consanguinity in the context of geographical locatedness. He explained that he was drawn to the field of genetics after working in a hospital in the Veluwe, a part of the Bible Belt:

> **KST:** What is it that attracted you to the field of genetics?
> **Bart:** Well, when I was in medical school—in Holland you have to work in a hospital as a junior assistant during medical school—I was working in Heidewijk in their pediatrics department, and there were a lot of inherited diseases because there was a lot of consanguinity in that area.
> **KST:** Oh, really?
> **Bart:** Yes. And there were a lot of people that were having to cope with genetic diseases. So I thought well, this is an interesting specialty.
> **KST:** How do you explain, or is there an explanation for, the level of consanguinity in that area?
> **Bart:** I think it is the, well, like all over the world, there are subgroups, subcultures, that are very closed off and there are religious subgroups there. There are matings between, well, they are not all consanguineous, but there are a lot of families that remain in the same area and marry there, and then there are consanguineous marriages.
> **KST:** And that is an area known for that?
> **Bart:** It is. It is like . . . there is in Holland a belt, we call it, of reli-

> gious people—in Zeeland in the south, that goes through here, through Amersfoort, and . . . these are very strict religious people . . . and they are very strict. We call them Zwarte Kousen because they wear black socks.

Diana's comments, quoted above, articulate widely held opinions about religious minorities as being traditional, unmodern, and backward in comparison to the mainstream, secular Dutch population. Hans, Saskia, and Bart, however, underscore the aesthetic and scientific value of the very practices that they, like Diana, perceive as backward. Thus, at the same time they distance themselves from the social world of the religious communities making up the Bible Belt, medical geneticists value these communities both aesthetically and as a resource for the production of scientific and medical knowledge.

During my fieldwork I encountered only one other context in which aesthetic judgments were made routinely: case demonstrations in the audit and the LOG.[11] These demonstrations were made up of presentations of complicated or unusual cases that had been resolved through diagnosis. Clinicians at the audit commonly reacted to the conclusion of a demonstration with a chorus of "beautiful diagnosis" (*mooie diagnosis*) or simply "beautiful." After a demonstration at the LOG that they found interesting, clinicians would frequently comment to each other, or to those sitting near them in the meeting, "Beautiful diagnosis."

An article in the *New York Times* in 1995 by the science reporter Natalie Angier suggests that in genetics, aesthetics and scientific knowledge are integrally linked. Writing about the German Nobel Prize–winning scientist Christiane Nüsslein-Volhard, Angier states,

> Within the last year or so, she [Nüsslein-Volhard] and her co-workers have been able to push into the next stage of the project, the real science —making mutants. That is what she has been reporting on lately, the 350 or so beautiful mutants that the laboratory has generated and begun to analyze. The power of genetics lies in just this process: if you want to understand how genes work in development, you try to muck up those genes and then see the results. . . . The hallmark of a topflight geneticist is the capacity to know a good mutant when she or he sees one: to detect the slightest deviation from normal development and to recognize that the deviation is significant, that it reveals something fundamental. (Angier 1995:C10)

These cases illustrate Pierre Bourdieu's argument that aesthetic judgments are historically specific and are made within specific and "highly autonomous field[s] of production"; he writes that such judgments are made in reference to history or information available to a specific group of people who have access to knowledge about a particular field. For Bourdieu aesthetic judgments are a function of power because they are formed within an autonomous field "capable of imposing its own norms on both the production and the consumption of its products" (Bourdieu 1984:3–4). Not only is the aesthetic value attributed to "beautiful chromosomes" specific to the field of genetics, but also, at least in the case of Hans's pleasure in chromosomes from the Bible Belt, the attribution is self-consciously so: Hans clearly linked his judgment of aesthetic value to his professional interests. He made a point of emphasizing the need to remember that people who supply beautiful chromosomes do not share his aesthetic judgment.

In her analysis of high energy physicists, Sharon Traweek discusses the contextual specificity of aesthetic values. She suggests that developing the ability to make appropriate aesthetic judgments is part of the professionalizing process involved in training young physicists. She says that during their graduate training physics students are "learning to become meticulous, patient, and persistent, and that these emotional qualities are crucial for doing good physics. They also are beginning to learn what is meant by 'good taste,' 'good judgment,' and 'creative work.' . . . They are receiving training in aesthetic judgments as well as in the emotional responses appropriate to those judgments. . . . They are learning to live and feel physics" (Traweek 1988:82). Training in making aesthetic judgments is an aspect of what is happening in the audit and the LOG. Understanding the historical and professional specificity of aesthetic values, however, does not illuminate what exactly creates the aesthetic charge the geneticists get from beautiful diagnoses, beautiful chromosomes, or beautiful mutants. Geneticists appear to generate aesthetic value as part of developing diagnoses and gaining professional recognition.

Geneticists at the clinic were delighted by what they interpreted as a successful resolution of a case—that is, a case concluded through diagnosis. Clinicians like Hans enjoyed the challenge of diagnosing complex syndromes. When I asked him if there were some types of cases he was interested in more than others, he told me he liked dealing with syndromes, skeletal cases, and metabolic disorders. I asked him why these types of cases intrigued him, and he responded,

> Especially, I think, because of the fact that [with syndromes] you have to puzzle over them, you have to look very carefully, you have to understand the gestalt well and compare it with the literature [on the disorder], listen carefully to the parents, [and think] hey, maybe that [something the parent describes] is a symptom, puzzling. I hate puzzling but I like puzzling with patients . . . it [is] very intriguing to make the diagnosis. A general principle in my choices, and in my work, [is that] I like to make diagnoses. Firm diagnoses and puzzling and comparing with the literature and then talking with parents.

Although some other physicians expressed less interest in the abstract nature of diagnosing complex syndromes, they pointed to their pleasure in the achievement of developing diagnoses. Saskia, one of the most junior physicians on staff, told me she preferred dealing with skeletal cases specifically because she found developing diagnoses in such cases more objective and less open to the need for interpretation than is the case with other syndromes.

Writing about beauty and physics, the physicist Anthony Zee suggests that aesthetic values are central to contemporary physics. He says that the "search for beauty is also one of the guiding principles of fundamental physics" (Zee 1992:815). He argues that for physicists, aesthetic value stems from symmetry and simplicity.[12] The pleasure physicians gain from developing a single explanation for the multiple and seemingly incoherent symptoms associated with various genetic disorders suggests that simplicity is a similarly significant component structuring their aesthetic sensibilities. In the case of medical genetics this is also an aesthetics of normalization and classification that, in stark contrast to a Calvinist view of monstrous births, can cast the abnormal as beautiful.

Both the local and the wider professional recognition physicians gain from producing diagnoses enhances their pleasure. Moreover, aesthetically pleasing diagnoses are likely to produce greater recognition. When a clinician presents a demonstration of the successful diagnosis of a case, she or he gains recognition from colleagues for her or his individual professional abilities. Recognition comes in the form of judgments about the beauty of the work that produced the diagnoses, judgments that help produce and reinforce the field in which aesthetic judgments are relevant. Professional recognition from the production of beautiful diagnoses can also extend beyond the immediate circle of geneticists at the clinic or even at the LOG. For example, Hans describes his patient from Meerwijk as being famous, but the patient is known, and Hans himself is recognized, because he published a scholarly article based on the case.

Nüsslein-Volhard's research focuses on the zebra fish as a means of understanding basic patterns of development. She produces mutant fish with genetic anomalies in laboratory tanks. Geneticists who work with humans do not deliberately produce beautiful mutants. Rather, if they want to understand how genes work they have to find people with existing genetic conditions whom they can follow and from whom they can establish "diagnosis and not guesses or maybes," as Hans put it. They find these beautiful diagnoses in communities such as those in the Bible Belt, where a preference for endogamy produces something like a natural laboratory. In such cases the aesthetic value comes not from phenotypically beautiful people but from the cells and their chromosomes, as extracted and abstracted from living human bodies.

RESOLVING THE ZEEDRECHT CASE

The clinicians regard the presence of a serious recessive disorder as exceedingly rare. "Most people," they repeatedly told me, "are healthy." Cystic fibrosis occurs in one out of every thirty-five hundred births (Cystic Fibrosis Foundation).[13] In the United States about 4 percent of the population, or one out of every twenty people, are unaffected carriers of the gene for cystic fibrosis (www.wrongdiagnosis.com). Although studies of European populations and populations of European descent have shown prevalence rates ranging from one in seventeen hundred births to one in sixty-five hundred births, a study in 1977 showed a prevalence rate of one in thirty-six hundred births, suggesting that one of every thirty Dutch people is an unaffected carrier of the gene. Cystic fibrosis is one of the most common recessive disorders—and the most common in European populations and populations of European descent. Such numbers mean that the chances of two separate recessive genetic conditions randomly occurring in one family are vanishingly rare. Indeed, as I argued in previous chapters, the explicit purpose of diagnosis is to produce a single explanation for a complex clinical picture in a particular individual. It is for this reason that senior clinicians emphasize the idea of "one child, one problem, most of the time." Nevertheless, it is a biological fact that the more closely related two people are to each other, the more genetic material they have in common. Geneticists estimate that every individual carries a small number of genes (2–4) that do not affect them but that would be lethal if they had two copies. Two closely related people are statistically more likely to share single copies of a higher number of the same genes than two more distantly related people. The social fact of Veronica's and Joop's possible consanguinity is what provides an opening for Bart to consider the extraordinary possibility that they might share genes for more than one serious genetic condition.

The potential presence of two recessive disorders demands an explanation beyond mere chance. In the case of the couple from Zeedrecht, Bart's suggestion of the presence of a second recessive disorder needed to be explained. Bart was convinced that the couple might be carriers of two recessive disorders. Given this possibility, he believed that if the couple went through the expensive, physically and emotionally painful process of attempting to have a child through preimplantation diagnosis who is free of cystic fibrosis, the child might nonetheless have a different serious, potentially fatal genetic disorder.

Bart's presentation highlights the process through which physicians working in the field of medical genetics come to conflate clinical, scientific, and popular knowledges in producing the medical diagnoses that have very real consequences both for the production of scientific knowledge and for patients seeking to use genetic information in making reproductive choices. Clinicians' understandings of the identities and practices of people in certain geographical areas inform their medical conclusions and influence the clinical options available to those people. In turn, the clinicians' conclusions help determine the pool of people that participate in such research projects as Maastricht's experimental program in preimplantation diagnosis and thereby condition the scientific knowledge produced in such contexts.

In the normal course of events, Bart's colleagues would disagree with his diagnosis of two recessive disorders. Such a situation is simply too unusual. Bart, therefore, needed to convince his colleagues that there were reasons to suspect such a rare event. In this case there appeared to be a lot of evidence against consanguinity: the couple said they were not related to each other, her father was from Amsterdam, and they had two different mutations for cystic fibrosis. If this couple were consanguineous, the geneticists would normally expect them to have mutations for cystic fibrosis that were "identical by descent." By reminding his colleagues that the couple came from Zeedrecht, Bart played upon locally shared beliefs about geography, religious practices, consanguinity, and biology to build support for his diagnosis of a second recessive disorder. His mobilization of these ideas reveals an interplay of geographical and biological imaginaries which together shaped his diagnosis and gave it greater authority in the eyes of his colleagues.

The geneticists emphasize both the scientific and the aesthetic value of the chromosomes they get from religious communities because these beautiful chromosomes compel the imagined biology of the body to speak. For physicians and researchers these chromosomes provide diagnoses and not guesses, thereby enabling the production of aesthetically compelling resolutions and of medical and scientific knowledge.

The complicated nature of Veronica's and Joop's case illustrates the complex interplay between social and biological facts related to geography, religion, marriage practices, the transmission of genes and chromosomes, and the clinical practice of genetics in the Netherlands. In the popular imagination urban secular Dutch people use the small population of Gereformeerden as a contrast against which they constitute themselves as modern subjects. People outside the clinic view Gereformeerden as backward, unmodern, and other. Inside the clinic, medical geneticists' perceptions of this group are nuanced by clinical interpretations of the social practice of endogamy and the resulting biology that such practices entail. Thus, medical geneticists explicitly see the Gereformeerd population as both backward and beautiful because their chromosomes are imagined as a resource for the production of scientific and medical knowledge and, by extension, professional status.

Chapter Five

BOVINE ABOMINATIONS

CONTESTING GENETIC TECHNOLOGIES

Sir Joshua Reynolds, an antagonist [of seventeenth-century Dutch art], and Eugène Fromentin, an enthusiast, met in their agreement that the Dutch produced a portrait of themselves and their country—its cows, landscape, clouds, towns, churches.

—Svetlana Alpers, *The Art of Describing: Dutch Art in the Seventeenth Century*

My examination of genetics and identity in the Netherlands here moves beyond individual clinical encounters to broader concerns about the relationships among science, technology, bodies, and the nation. I want to examine how people in the Netherlands attempted to come to terms with the phenomenon of a transgenic bull, that is, a bull with a gene from another species (in this case a human gene) incorporated into its genome. My analysis focuses on how one group, the Dierenbescherming (the Dutch Society for the Protection of Animals), mounted a national campaign against genetic manipulation in animals as represented by the transgenic bull known as Stier Herman (Herman the Bull).

During my fieldwork I discovered a startling Madonna-like image in which a blond-haired (and presumably blue-eyed and therefore quintessentially Dutch) woman breast-feeding a baby has breasts that are cow udders (fig. 6). This deliberately unsettling poster, measuring four feet by five and a half feet, was placed inside the Plexiglas advertising space in bus stop shelters throughout the country. It was created by the Dierenbescherm-

6. "New! Mother's Milk from Cows!"

ing as part of its campaign against genetic manipulation in animals. The image was also reproduced in newspaper articles about the poster and about the Dierenbescherming's campaign. When I visited the organization's main offices in The Hague to talk to the staff about their campaign, they showed me a second poster, this one featuring the image of a cow staring straight into the face of the viewer; above the image was the text, "Soon with Blond Hair and Blue Eyes?" (fig. 7).

These images caused me to ponder what they could tell me about the relationship between genetics and culture. Marilyn Strathern suggests that culture itself lies in the processes through which people make connections between the various ideas available to them in a given context:

> *How* . . . connections are constructed, the way facts and opinions are brought together, reveal ideas in context. . . . Anthropologists . . . would say that culture lies in the manner in which connections are made,

7. "Soon with Blond Hair and Blue Eyes?"

> and thus in the range of contexts through which people collect their thoughts. . . . In so far as ideas are inevitably contextualised, that is, shaped through the forms in which they are expressed, all human activity has a generic cultural dimension. At the same time, cultural repertoires are differentiated from one another by the degree to which such forms appear distinctive; specific cultures afford people specific kinds of contexts. (Strathern 1993:4–6)

Although conceptions of culture abound in anthropology and elsewhere, Strathern's conceptualization is useful for thinking through the ways people transform a specific mode of knowledge, in this case genetics, which is presumed by most to be universal, even as they incorporate it into local contexts.

An analysis of the Dierenbescherming posters offers a means of understanding and illuminating broader issues concerning the production, interpretation, and practice of genetics in the Netherlands today. The posters point our attention to, among other things, tensions between genetics and contemporary conceptions of Dutch identity in the Netherlands. In so doing, they shed light on the simultaneous production of science and society in and through conceptions of nature and culture. The focus on cows can hardly be viewed as accidental. Cows are ubiquitous in the landscape and have enormous cultural resonance as symbols of Dutch society.[1] Manipulating the genetic "identity" of cows, therefore, may be seen as tampering with Dutch identity itself.

Moreover, the explicit reference the posters make to genetic manipulation and their implicit allusion to eugenics also arouse Dutch memories of the Second World War and antipathy toward Nazi science. Dutch interpretations of the legacy of the war continue to be powerfully salient in the Netherlands today. Popular interpretations of Nazi eugenic practices aimed at producing an alleged super race are pertinent in relation to genetics.[2] Most Dutch people consider Nazi eugenics to be the complete antithesis of the highly valued Dutch social ideal of tolerance.[3] The poster images therefore invoke the ongoing desire of most Dutch people to distinguish themselves from what they continue to perceive as their less tolerant German neighbors by drawing on broadly held concerns about genetics, borders, and the body as a means of publicly questioning and protesting genetic practices in the Netherlands.

These contestations further help one understand that genetic knowledge and practices cannot be reduced to biology, nor can they be understood simply as scientific knowledge and their application as a set of technological and medical practices. I explore how new genetic knowledge and associated technologies raise issues regarding purity and the porous border crossings and policings of contemporary biotechnoscience. The distinctive ways people in the Netherlands bring ideas together to make sense of a genetically anomalous entity further elucidates the levels of enculturation surrounding the production, consumption, and use of new genetic knowledge.

CONCEPTUALIZING GENETICS IN THE NETHERLANDS

The Dierenbescherming's campaign against genetic manipulation in animals raised questions about boundaries and national purity similar to those raised by the idea of a genetic passport. As Donna Haraway has pointed out, activists opposing the production of transgenic organisms "appeal to notions such as the integrity of natural kinds" (1997:60), but she warns of "the unintended tones of fear of the alien and suspicion of the mixed" (1997:61) in the articulation of such opposition. The images in the Dierenbescherming's posters as well as the idea of the genetic passport evoke and exploit precisely these kinds of issues. A passport explicitly references national identity and national borders and therefore, by extension, race as well (Balibar 1991 [1988]:37; Rose and Novas 2005). A genetic passport, by referencing DNA, implicitly refers to concerns about body boundaries and therefore individual and national integrity as well as race.

The significance of the idea of a genetic passport stemmed in part from its association with the history of the Second World War and the fact that, in contrast to many European countries, the only time Dutch people had been required to carry identity cards was during the Nazi occupation. Many Dutch people I spoke with considered the obligation to carry an identity card an infringement of their personal autonomy and linked the requirement to the intrusions and indignities of Nazi occupation. Others, however, stated that they were not concerned about the implementation of a national identity card system, as they had "nothing to worry about." Such expressions implied that those who did have something to worry about perhaps did not belong, thereby linking the discussion of identity cards to contemporary anxieties about increasing immigration associated with European unification. Likewise, the images in the Dierenbescherming's posters both explicitly and implicitly elicit issues of the transgression of body and national boundaries, the moral problems of genetics associated with the Second World War, and national and individual identity.

As I have elaborated in previous chapters, people frequently linked the services offered by the genetic centers (genetic counseling, pre- and postnatal genetic testing, and, increasingly, other genetic practices like preimplantation diagnosis) to a positive conception of Dutch identity as modern. Scientific and medical explanations for various genetic conditions available from genetics experts were preferred by many over other kinds of explanations—such as a "sign from God"—that they perceived as traditional and premodern.

However, the Dierenbescherming's posters and the controversies that followed highlight the tensions surrounding this conception of Dutch identity by linking genetic manipulation to this positive conception of Dutchness and modernity. I first encountered the poster of the woman with cow udders when a Dutch friend called my attention to the early news coverage on it. Whereas the genetic passport existed only as a well-elaborated myth that nonetheless possessed a social force that anthropologists fully recognize, Stier Herman was in fact a living organism. The practices that created Stier Herman are increasingly becoming the everyday practices of biotechnoscience, in which, for example, the production of human-animal chimeras—organisms containing both human and animal genetic material—has become an important tool in stem cell research. As I explored the controversy surrounding Stier Herman, I found a widespread and multilayered engagement with the issues his existence raised.

Anthropologists have long been attentive to the powerful roles animals play in human lives in those places Akhil Gupta and James Ferguson (1997) describe as the traditional sites of anthropological interest. Only recently have anthropologists turned their attention to the significance of animals in the wider range of contexts in which they now work. In inquiring into the relationship between Louis Henry Morgan's parallel interests in humans and beavers, Gillian Feeley-Harnik reiterates Ralph Bulmer's (1967) argument that "human beings project onto animals their conceptions of their own social relations" (1999:221). Referencing T. O. Beidelman (1980) and Judith Goldstein (1995), Feeley-Harnik writes, "Human beings are most apt to use animal imagery in dealing with moral dilemmas that are sensitive, difficult, or completely insoluble" (1999:221). Perhaps nowhere do taxonomic anxieties appear more insoluble than in contemporary biotechnoscientific interventions into what Sarah Franklin (2000) describes as "life itself."

Franklin argues that on the cusp of the twenty-first century the world is witnessing "a process of cultural redefinition whereby foundational understandings of the human, the body, reproduction, and the future are being transformed" (2000:188; also see Haraway 1997; Strathern 1992a, 1992b). She suggests that the "removal of the genomes of plants, animals, and humans from the template of natural history that once secured their borders, and their reanimation as forms of corporate capital" opens a "broad Foucauldian question of how [these phenomena] can be understood as part of an ongoing realignment of life, labour, and language" (Franklin 2000). Given their historical attention to the diversity of foundational concepts that can be found in the world, anthropologists are well situated to attend to both the localizing and the boundary-crossing processes at work as people attempt to make sense of biotechnoscientific practices that appear to shake foundational understandings of the human, the body, reproduction, the nature of nature, and life itself. These dynamics are powerfully at work in the Dierenbescherming's campaign, in which the use of animals played a central role in expressing intense anxieties about emerging genetic technologies. By associating monstrous animal-human hybrids, Nazi science, and the border crossings these entail, the campaign suggested that such technologies were potentially as morally insoluble as the history of the Second World War and the problem of contemporary immigration.

INTERPRETING THE DIERENBESCHERMING CAMPAIGN

Founded in 1864, the Dierenbescherming is well established and widely known as one of those civic organizations that constitute the consociality of Dutch public life (Kruijt 1974; Post 1989; Shetter 1987). Despite the radical challenge presented by their posters contesting the genomic manipulation of dairy cattle, the Dierenbescherming is a thoroughly mainstream organization, analogous to the Humane Society in England and the United States.[4] Founded in The Hague by Dutch business elites and endorsed by King Willem II, it was organized around what were termed the uncivilized and improper abuse of animals in activities such as dog racing. Early on, the organization also became involved in protecting animals used in agricultural production, such as cows and horses. The organization quickly spread to the national level, opening local chapters throughout the Netherlands, and now has over two hundred thousand members.[5]

During my fieldwork, the Dierenbescherming was in the middle of a campaign against genetic manipulation in animals. Although aimed at agricultural genetics and articulated in these terms, the Dierenbescherming exploited images that engaged directly with lingering attitudes about Nazi science and with new anxieties about purity in light of increasing immigration and European unification. The Dierenbescherming's campaign targeted genetic manipulation in all animals, arguing that animals should be cherished for their intrinsic value and not turned into "vessel[s] in which medicinal substances are produced by means of chemical reactions" (Dierenbescherming n.d.). They argued that people are "on the verge of extinguishing the original animals and replacing them with living machines" (Dierenbescherming n.d.).[6]

The most striking element of the public education phase of the Dierenbescherming's campaign against the use of biotechnology in animals was the poster of a woman with cow udder breasts. The text of the poster reads:

> *NIEUW* Moedermelk uit koeien! STOP BIOTECHNOLOGIE BIJ DIEREN. Koeien met menselijke genen moeten moedermelk gaan geven. Genetisch manipulatie wordt onze kinderen met de paplepel ingegoten. Help de Dierenbescherming deze waanzin te stoppen!
>
> [*NEW* Mother's milk from cows! STOP BIOTECHNOLOGY IN ANIMALS. Cows with human genes have to produce mother's milk. Our children are being fed genetic manipulation. Help The Dierenbescherming stop this insanity!]

The text then asks the reader to become a member of Dierenbescherming or ask for more information.

Translation is always complicated. The sentence I have translated as "Our children are being fed biotechnology" is highly colloquial and therefore difficult to translate. The part of the sentence I have translated as "our children are being fed" may not completely capture the nuance of the expression in Dutch—"wordt onze kinderen met de paplepel ingegoten." The reference to *paplepel* is to a "pap" spoon or baby spoon. The expression therefore directly indexes children being "fed" biotechnology, or genetically manipulated products, with their baby food. Indeed, the expression could well be translated as "Our children are sucking in genetic manipulation with their mother's milk." Such a translation would be confusing in this context since the concern expressed by the poster is that Dutch children will not be sucking at their mother's breasts at all but on bottles containing milk from genetically manipulated cows, animals born of biotechnology. In settling on the translation I use here, I have attempted both to capture the meaning of the poster's text and to convey the succinctness of a campaign poster slogan.

Stier Herman, the transgenic bull whose existence inspired the Dierenbescherming's campaign, was laughingly described to me by a number of Dutch people as "our most famous citizen" (fig. 8). Comments like these, in which cattle are recognized as citizens, point to the potent symbolic status they have in the Netherlands. A transgenic animal is one that has a gene from another animal or organism incorporated into its germline DNA. Inserting a gene into an animal's germline DNA means that it passes the gene on to its offspring. Stier Herman, the world's first transgenic bull (Boer et al. 1993), was developed by the biotechnology company Gene Pharming Europe and its American parent company, GenPharm International. Herman has a human gene inserted into his genome so that his

8. Stier Herman with his first female offspring.

female offspring will produce milk containing the human form of the protein lactoferrine. All lactating animals produce a form of lactoferrine. The human form of the protein acts to prevent udder inflammation in cows and can be used in medicines for fighting intestinal infections in humans (Köhler 1991). The text in the poster about mother's milk from cows refers to the fact that the enzyme structure of milk from Herman's female offspring makes their milk more similar to human milk than milk from other cows.

Stier Herman, of course, is not a cow, but a bull, and there is, after all, a difference. That the Dierenbescherming's posters figure national anxiety on the female body can hardly be viewed as neutral. Since at least the 1960s, there has been renewed focus in industrialized countries on breast-feeding as the royal road to a natural relationship to one's child and to the healthy development of the child's psyche and immune system. Thus, like genetic counseling, breast-feeding figures as a contemporary strategy for successful reproduction. Additionally, women, breasts, purity, and milk have long been interpolated into ideas and practices concerning nationalism, colonialism, modernity, nature, race, and eugenics. Feminist scholarship on these issues has pointed to the ways women and their bodies historically and in the present do important ideological work in creating and maintaining what are in fact porous social boundaries (Haraway 1991, 1997; Martin 1987; McClintock 1995; Poovey 1988; Schiebinger 1993; Stoler 2002; see also Kevles 1995 [1985] and Proctor 1988). The Dierenbescherming's posters perform similar ideological work by illustrating how biotechnologies associated with modernity may transgress species boundaries between human and animal. Specifically, they present a transfigured female body, foregrounding the woman's breasts, thereby emphasizing that the transgression not only occurs on the body of the woman but also will be passed on to her children—the next generation of the Dutch nation—through milk. By focusing on the female body, the posters offer a critique not only of how biotechnology threatens to transform the individual, but also the identity, health, and purity of the Dutch nation itself.

The legal guidelines surrounding biotechnology and animals in the Netherlands are based on a "no, except . . . principle" (*nee, tenzij . . . principe*) (Dierenbescherming 1992:9). The no, except . . . principle means that biotechnology involving animals is illegal in the Netherlands except under guidelines developed by the Ministry of Agriculture, Nature, and Fisheries, as advised by the Department for Biotechnology of the Council for Animal Affairs (De Afdeling biotechnologische aangelegenheden van de Raad voor dieraangelegenheden) (Dierenbescherming 1992:9). The legal concept of a no, except . . . principle seems uniquely suited to Dutch goals of accommodating diversity in the face of controversial issues because it recognizes the possibility of difference within the Netherlands and, as I elaborated in chapter 1, the need to accommodate such difference.[7] It also suggests the assertion of a national difference having to do with establishing the sovereignty of the state to delegate authority over these kinds of issues to an agency that will both articulate and sustain Dutch ideas of what might be desirable.[8] The same no, except . . . principle guides Dutch policy on euthanasia. In both cases, formal prohibitions acknowledge the legitimacy of the position of those for whom activities such as genetic manipulation and euthanasia may be morally repugnant. At the same time, the guidelines allow that genetic manipulation and euthanasia might be appropriate, even desirable, under certain circumstances. The Dutch state thereby effectively delegates authority to experts whose special knowledge becomes the basis for managing difference of opinion and developing guidelines for appropriate action.[9] The guidelines enabled the production of Herman because one of the exceptions to the prohibitions was in cases in which the biotechnology would lead to products valuable for human health that could not be effectively and efficiently procured in another manner. Gene Pharming Europe claimed that the production of lactoferrine in this way met these criteria.

The significance of this exception is contextualized by the desire of the Dutch government to make the Netherlands a desirable place for locating biotechnology businesses. In discussing Herman with me in an interview, a pharmaceutical industry employee mentioned that the government had identified biotechnology as an industry that it would like to attract to the Netherlands. Creating the social and political conditions through which scientific practices associated with biotechnology would be supported would obviously be an important component of attracting this industry.

As if to underline the tensions evoked by the Dierenbescherming posters, a political scandal involving Stier Herman erupted while I was in the Netherlands; the scandal greatly augmented the notoriety of the poster of the woman breast-feeding from cow udders. The explicit message of the poster is that milk from Herman's female offspring might come to replace mother's milk in the feeding of Dutch infants. Gene Pharming Europe had steadfastly insisted that its project was aimed strictly at developing medicines and reducing udder inflammation in cows and that the project was not meant to replace breast-feeding. In June of 1994, the press began to report that from 1990 to 1993 Nutricia, a primary producer of baby formula in the Netherlands, had supported Gene Pharming Europe's research on Herman. According to press reports, the contract between Gene Pharming and Nutricia gave Nutricia marketing licenses for proteins from the milk from transgenic cattle (Köhler 1994:1). This revelation came in the midst of the Dierenbescherming's public education campaign involving the poster of the woman with cow udder breasts. The scope of the scandal was considered serious enough that members raised it on the floor of parliament, demanding to know why the Dierenbescherming seemed to have a better picture of what was happening with Stier Herman than members of parliament.

In the context of rapidly expanding biotechnoscientific practices, yesterday's abomination often becomes today's ordinary practice. This did not happen in the case of Stier Herman in the Netherlands. While globally the technologies associated with Stier Herman have become increasingly common, in 1996 the production of transgenic cows in the Netherlands was halted because the levels of lactoferrine in Herman's offspring were so low that the project did not meet the guidelines for exception from the prohibitions against genetic manipulation in animals. Gene Pharming closed its Dutch farm and moved most of its transgenic cattle to Finland in 1998. They left behind only Stier Herman and Holly and Belle, two cloned cows they had subsequently produced. Herman, Holly, and Belle were moved to Naturlis, a natural history museum in Leiden, where they were supported by contributions from Yarden, an organization focusing on end-of-life issues, and Nedap, a diversified Netherlands-based multinational corporation whose businesses include antishoplifting technology, information processing, election systems, health care, and cattle management. Holly and Belle are still on exhibit, but on 2 April 2004 Herman was euthanized because he was suffering from arthritis (Associated Press 2004).

In 1993 the Dierenbescherming posters cast the issue of genetics in terms of social policy, calling on citizens to join their organization with the intention of strengthening the group and its ability to lobby parliament. This politicization of genetics challenged the legitimizing discourses employed by geneticists to cast genetics as a matter of individual choice. By locating choice at the societal level, the posters challenged and problematized the idea of choice exercised at the individual level when it comes to issues about genetics. The posters insisted on situating individual choice in the broader society beyond the confines of places such as clinics and laboratories. Indeed, the implicit message of the posters is that individual choice becomes meaningful only in the context of larger social, historical, and political issues.

A further message of the posters and the scandal they provoked is that society as a whole has neither adequate control over the agenda of biotechnology nor meaningful choice regarding how it may enter their daily lives, their bodies, and the nation. The text accompanying the poster of the woman with cow udder breasts literally speaks of biotechnology entering the body at a formative stage through milk. The posters actually argue for limiting individual choice by banning such uses of biotechnology.

Anthropologists such as Mary Douglas (1966), Victor Turner (1995 [1969]), and Emily Martin (1990b, 1991) have written about the power and significance of bodily fluids like milk. In his classic analysis of ritual and symbol, Turner points out that the power of a symbol derives from the way it condenses and unifies multiple referents. He argues, "A single symbol, in fact, represents many things at the same time: it is multivocal, not univocal. Its referents are not all of the same logical order but are drawn from many domains of social experience and ethical evaluation. Finally, its referents tend to cluster around opposite semantic poles. At one pole the referents are to social and moral facts, at the other, to physiological facts" (Turner 1995 [1969]:52).

Cows are multiply symbolic, as Dutch icons and as producers of milk and other dairy products that are major staples of the Dutch diet. Thus, at the level of social and moral facts, it is no coincidence that the Dierenbescherming's challenge to biotechnology was framed around images of cows, a quintessentially Dutch animal. As I mentioned above, some Dutch people readily consider cows as citizens. Along with windmills and tulips, cows have long served as a central icon of the Dutch nation. Cows are everywhere in celebrated works of Dutch art (Alpers 1983). The symbolic significance of cows is also apparent in the way people think about the landscape. When I asked people in interviews how I would know I was not in the Netherlands anymore, one of the most common responses was that the landscape would be different. They described the Dutch landscape as one of flat green meadows with (mainly black and white) cows grazing on them. Once while bike riding past a field of corn a friend told me that such a scene was new: fields of corn were a new addition to the Dutch landscape. Another friend made a similar observation, telling me that one could frequently hear expressions of concern about the increase in corn production in the Netherlands out of worry that the traditional Dutch landscape of meadows and cows would disappear. She said she did not think such a transformation would ever occur because much of the land in the Netherlands is simply too wet to support anything but grass. Cows have also gained symbolic significance in popular media. For example, the Dutch film awards, the equivalent of American Oscars, are known as Golden Calves (Gouden Kalveren).

At the level of physiology, in the Netherlands consuming dairy products is practically a national pastime. In this context milk is an especially potent symbol of the constitutive elements of the Dutch body and Dutch social relationships. In discussing the plethora of Dutch dairy products with me, one informant suggested that new products had to be continually invented to help sell the huge quantities of milk produced by the large population of Dutch cows. Another discussed problems associated with the large number of cows in the Netherlands in a rather comic way. He explained that there were starting to be problems with groundwater in the Netherlands because of the large volume of cow dung in the country. He laughed and said, "The country is filled with entirely too much shit!" One could argue that the Dutch are preoccupied with their cattle, and that the symbolic significance of cows reflects a bovine idiom at work in the Netherlands (Evans-Pritchard 1940). It is precisely this preoccupation with cows that the Dierenbescherming's campaign makes use of. The campaign uses cows to project and refract deep anxieties about the moral problems new genetic technologies pose for contemporary Dutch social relations.

In showing how biotechnology distorts the integrity of cows, the Dierenbescherming posters also suggest that biotechnology distorts the integrity of Dutch national identity. Furthermore, the woman/cow image reads as a quintessentially Dutch woman with blond hair and blue eyes. The transgression of her body boundaries thereby signals the potential transgression of the identity, health, and purity of the Dutch nation. The images on the Dierenbescherming posters gain special force from the ways the organization has constructed them. Animals function as "boundaries and can come threateningly, or alluringly, close to humanity" (Lindee, Goodman, and Heath 2003:9). The conceptual problems posed by this phenomenon in relation to genetics are formidable because of how genetic knowledge and practices, especially the production of transgenic organisms, emphasize the relatedness of all life forms (Marks 2002). In the posters, the Dierenbescherming has manipulated bodies of women and cows to construct images signaling danger and pollution because they disrupt a longstanding, socially valued classificatory distinction between human and nonhuman. In so doing they illuminate the complex multiple ways such schemes are both embedded in and produced through a diverse array of social processes, including historical consciousness, national identity, gender, understandings of the body and bodily fluids, social relationships, scientific and medical research, international biotechnology, and local social values regarding animals, humans, and genetics.

TRANSGRESSING BOUNDARIES

The posters transgress social, ethical, historical, and political boundaries that work to articulate distinctions between nature and culture. One poster represents a jarring image that is neither fully human nor fully animal made through the fusion of nature and culture. This grotesque hybrid, at once fascinating and repellent, explicitly transgresses body boundaries.

In so doing, the poster and the biotechnology it challenges disrupt longstanding classificatory schemes distinguishing human from nonhuman by invoking their other—the chimera, the hybrid, the monster. These figures have a long history in European culture (Davis 1975; Stewart 1993) and serve as an important trope in technoscientific imagery (Haraway 1991, 1997). Susan Stewart argues that anthropological studies of symbolic inversions, including those associated with the grotesque body, have emphasized how such practices reaffirm important cultural categories (Stewart 1993:106). In discussing the importance of classificatory schemes, Mary Douglas (1966) argues that such taxonomies serve to maintain social order and that their disruption simultaneously signals danger and desire. Stewart wants to argue that the trickster upon which such studies often rely is "also a spirit of creativity, a refuser of rigid systems" (Stewart 1993:106). In this sense, Stewart argues, transgressive practices serve both to create culture and to challenge cultural norms (Stewart 1993:106). Understandably disturbing to anyone who is able to read the image, the poster is especially threatening in a society such as the Netherlands that relies so heavily on categorization: first, to maintain social and ontological order and, second, to reproduce itself as a self-consciously pluralistic, tolerant society.

The threat to boundaries conveyed by the poster images gains force in Dutch society by their invocation of Nazi science. As Nazi genetic experimentation crossed an ethical boundary between what is today considered legitimate and illegitimate science, the poster delegitimizes contemporary biotechnology by characterizing it as an insane manipulation of human and animal bodies alike.[10] The reference to Nazi science is salient in the Netherlands because of the way it serves to represent the complete antithesis of the powerful Dutch social ideal of tolerance. Although the reference to Nazi science is mostly implicit in the poster of the woman/cow hybrid, the Dierenbescherming makes the connection far more explicit in the cow poster. While not grotesque, this second poster explicitly links Nazi science to contemporary efforts at genetic manipulation. The image's suggestion of the possibility of creating an "Aryan" race of cows speaks directly to popular understandings of Dutch memories of Nazi attempts to create a genetically engineered super race.

People I encountered both inside and outside the genetics centers repeatedly made connections between genetics and Nazi efforts to produce a super race as part of its program of racial hygiene.[11] In interviews, people stressed the importance of drawing and maintaining boundaries between appropriate uses of genetic technologies and inappropriate uses like those made by the Nazis. They used boundary metaphors with reference to Nazi science most frequently in discussing prenatal testing and Stier Herman. When I discussed Stier Herman with Tom, a forty-year-old public health official whose child has a relatively unusual genetic condition, he told me that the purpose of developing Herman was for public health (*volksgezondheid*). But he went on immediately to talk about how Herman raised the specter of science fiction images of science gone wrong, such as in the film *The Boys From Brazil* (a film about cloning Hitler), and memories of the Second World War. He said that "the limit has to be set" (*de grenzen aan moet stellen*), there have to be "strict rules."[12]

A couple in their late forties who also have a child with a rare genetic condition also spoke with me about Stier Herman. When I asked them if they had heard of the transgenic bull and what they thought about him, they responded, "Where is the limit [*grens*]? We cannot understand where it is going. The idea of a super race is a very bad business." They went on to talk about the expense of a well-organized social welfare system and then returned to the idea of the production of a super race: "No one is perfect, and in my opinion that is firm. It is a creepy idea that begins with superhumans. You are as you are."

Margot is a forty-two-year-old pharmacist's assistant whose eldest child has cystic fibrosis. In discussing prenatal testing with me, she too raised the issue of limits. When I asked her opinion of prenatal tests such as amniocentesis she said it is "good that one can do it, but I am glad that I did not have to make a choice. [It's a question of] limits. I think it is fine if it is for illnesses but if it is to get a superchild, I am against that."

Finally, Peter, a thirty-eight-year-old office worker whom I first met when he came to the prenatal outpatient clinic with his wife for genetic counseling, also invoked the problem of Nazi science to distinguish between the potential use and misuse of techniques such as those employed in producing Stier Herman. He said that developing a bull like Herman was "a logical path. But, I am a bit anxious about what will happen in the future. I am for it and against it. It is a bit creepy because it raises the issue of creating a super race." Peter's ambivalence captures the tensions between popular hopes for modernity realized and apprehensions over modernity gone awry. Taken together, these examples highlight the ways people express their anxieties about violating boundaries in talking about genetics, suggesting that they see in Stier Herman the limit of acceptably transformed nature.

Widespread understanding of a shared experience of suffering and oppression during the war serves as an important reference for articulation of a distinctive Dutch identity, shared, however unevenly, by geneticists and laypeople alike (van de Braak 1993; Shetter 1987; van der Zee 1982).[13] Popular Dutch perceptions of Nazism are linked to an awareness of the Nazis' aims to produce an Aryan race through eugenic policies. These policies, in turn, are widely perceived as being deeply antithetical to the Dutch social ideal of tolerance. As a result, contemporary genetic practices in the country have been and continue to be forged in relation to the legacy of Nazism and German occupation. Many Dutch people use the legacy of the war to articulate the connections they use to distinguish themselves from their German neighbors. This is a long-standing, ongoing process involved in the forging of Dutch identity that is made both difficult and necessary by the many similarities between Dutch and German people (Barth 1969). It gains added significance today in the context of the ongoing move toward European unification.

The ways Dutch people continually reference these issues in distinguishing themselves from their German neighbors may not be obvious to outsiders. The link between genetics and the Second World War implicit in the concept of the genetic passport, for example, is apparent only to those who know something of the history of identity cards in the Netherlands. The Dierenbescherming's campaign and posters similarly articulate the distinctiveness of Dutch in relation to German. They do so, in part, by speaking to the contrast between the Nazi aim of producing a super race and the Dutch social ideal of tolerance.

Over the course of my fieldwork I came to recognize the ideal of tolerance inhering in every conception of Dutch identity that I encountered. I came to understand this Dutch ideal as involving not only tolerance, but also the acceptance, understanding, and valuing of difference, linked to the Netherlands' long history of religious tolerance, historically managed by the social structure of *verzuiling* (pillarization).[14] As the previous chapters illustrate, the lasting legacy of the social structure of pillarization enables Dutch people to recognize significant social differences without necessarily stigmatizing them; the structure allows people to bound and contain difference within known and accepted categories. That achieving the ideal of Dutch tolerance relies so heavily on categorization helps explain why category confusions such as those represented in the Dierenbescherming poster of a woman with cow udder breasts could be perceived as a threat to the integrity of Dutch national identity, the Dutch nation, and social order.

Dutch people continually enact values around the ideal of tolerance in living their daily lives. For example, during my fieldwork I soon realized that Dutch people are quite reluctant to label others as intolerant. Such a label would be a serious insult and a sign of their own intolerance of difference. Rather, in interviews people frequently told me that Dutch people differ from their neighbors by describing German and French people as a little or a bit "more chauvinistic," a concept less socially charged (and hence more polite) than that of intolerance. Dutch interpretations of the Nazi goal of producing a super race through its eugenic program of racial hygiene afford a particularly well-elaborated example of German intolerance. Because Nazi eugenics represents an affront to the ideal of tolerance, Dutch geneticists never cease in their efforts to distinguish contemporary genetic practices from those associated with the Nazis. The combination of Dutch interpretations of the Nazi legacy, the social importance of tolerance, and the ongoing process of distinguishing Dutch from German converge in powerful ways to contextualize genetics in the Netherlands today. This context helps account for the readiness with which Dutch people draw on the Nazi example to discuss the risk to national and personal integrity posed by genetics.

The political salience of the challenge to boundaries presented by genetic research in the Netherlands is further evidenced in a report titled *Genen en Grenzen* (Genes and Boundaries) (CDA 1992) published by the Dutch Christian Democratic Party, which governed the Netherlands from the end of the Second World War until 1994. The report attempts to delineate the party's official view of the boundary between acceptable and unacceptable applications of genetics. That the Dutch people I met found concepts about boundaries and limits so useful in talking about their hopes for and fears of contemporary genetics underscores the perception many Dutch people have about the potentially transgressive nature of this scientific and medical practice. These hopes and fears, though variously interpreted, are widely shared by people inside and outside the clinics.

The focus on setting limits helps point to an underlying distinction many Dutch people seek to make between genetics as appropriate modern medicine and genetics as a sign of modernity gone awry, as evidenced for them by Nazi eugenics. Genetic practices pose social and conceptual problems in the Netherlands when they are aimed at transgressing categories of normal by producing animals or humans that people perceive as being somehow supernormal or impure. Both animals and humans are considered to have inherent value. People widely accept genetic counseling insofar as it respects the inherent value of humans by respecting the boundaries of normal and ordinary humans. However, they challenge the genetic manipulation of humans and animals because this transgresses the boundary of what they consider to be normal and ordinary. Although I have included only a few examples here, people raised the issue of boundaries and limits in virtually all of the interviews I conducted.

In *Beter dan God*, the television documentary from 1987 in which Huub Schellekens introduced the concept of a genetic passport, a number of other people raised the history of Nazi science and the possible link between contemporary genetics and past eugenic practices.[15] Dr. Gooren, an endocrinologist who works on impotence and infertility, discussed this potential in the film when he suggested that with the development of human genetics people will slowly begin to "eliminate people who differ from the norm." Wim Kayzer, who made the film and is interviewing Gooren, says, "I think that if people sit and watch [this program], they will think, 'now, Mr. Gooren, that is good, but that would [not happen] in the Netherlands, a nice constitutional state, parliamentary democracy . . . we would all really neatly stop that.'" Gooren then responds, "Then I think that people in the Netherlands forget. It will, of course, not happen all at once, it will go millimeter by millimeter. I think then, for example, of Nazi Germany. That is, of course, not a flattering comparison with the Netherlands, but there the most horrible things also did not happen in one night. Things went millimeter by millimeter. People accustom themselves to certain things. If ten years ago we had spoken about test-tube babies, then I would have said, everyone would have said, 'that is crazy.' But now it is an ordinary thing" (Kayzer 1987:18). These examples indicate how scientific authority is vulnerable to popular identification with Nazi science and with Germany. It is precisely this vulnerability that the Dierenbescherming posters exploit.

The above examples illustrate how readily Dutch people invoke the history of Nazi eugenics in talking about contemporary genetic practices. The readiness with which many Dutch people use the Nazi case, however, is not just about distinguishing potentially "good" from "bad" genetics. Rather, what the Dutch people I met seemed to be trying to work out is how to use new biotechnoscientific knowledge and practices that challenge previously held understandings of the structure of the natural and social world in an appropriately Dutch way. The evocation of Nazi eugenics offers people an opportunity to distinguish the Netherlands and its citizens from their larger, more powerful neighbors. This is an ongoing process with deep historical roots for which the Second World War and Nazi atrocities and occupation offer an available set of shared references for articulating Dutchness. If one takes seriously Balibar's claim that the "discourses of race and nation are never very far apart" (Balibar 1991 [1988]), then one must understand the processes I am describing as articulating Dutchness not only as not German but also as not anything else either. In this case, then, the Dierenbescherming posters depict—and offer Dutch people an opportunity to claim—a traditional image of Dutchness and the Dutch nation in a country with increasingly open borders that is now and ever more *vol op* (full), not just of Dutch people but also of relatively recent immigrants.[16]

The Netherlands is frequently identified as having the highest population density in the developed world. *Nederland is vol op* is a phrase used to express the idea that the Netherlands cannot absorb more people. Indeed, one might view the genetic passport's suggestion that there are people in the Netherlands who do not belong there as foreshadowing more recent (and more hostile) sentiment in this regard. During my fieldwork I certainly heard Dutch citizens voice concerns about the presence of immigrant communities, particularly those made up of Turkish and Moroccan immigrants and their families who entered the country as so-called guest workers in the 1970s. Such issues were often expressed in relation to the idea that the immigrants were not learning culturally salient practices such as speaking Dutch and riding a bike. There was, nevertheless, also a national conversation about the Netherlands now being a multicultural society (something that teenagers in particular articulated in conversations with me) and an explicit commitment to recent immigrants that included, for example, teaching immigrant children in their own languages in primary school and providing state support for newspapers and radio and television programming that reflected and expressed the dominant views of the largest immigrant populations.[17] But as discussed in chapter 1, by

2002 a new anti-immigrant political party had achieved significant success, and new social policies reflected a shift away from these attitudes.

LEGITIMATING DUTCH GENETICS

Dutch geneticists repeatedly told me that they are very aware of the power of the identification popularly made between genetics and Nazi science to undermine the legitimacy of their work. During my fieldwork, I found them continually engaged in distinguishing contemporary genetics from practices associated with Nazi science. Evelyn Fox Keller has pointed out that the Nazi legacy significantly influenced the interpretation and practice of genetics in the post–Second World War period, stating that "in the revulsion against Nazi eugenics in Germany . . . the direct link between genetics and its eugenic implications . . . was no longer politically tolerable" (1992:285). Geneticists in the postwar era had to reinvigorate their discipline by distinguishing their practices from Nazi eugenics. They did so, first, by demarcating knowledge from the potential uses of that knowledge—the old argument of use versus misuse of scientific knowledge—and, second, by confining genetics to the interpretation of "purely physiological attributes," leaving behavior to the realm of culture (Keller 1992:285).[18]

Interestingly, Keller's tracing of the history of genetics is in response to her belief that there is a contemporary resurgence of the link between genetics and an increasingly wide array of human attributes: "Today we are being told—and . . . we are apparently coming to believe—that what makes us human is our genes. Indeed, the very notion of 'culture' as distinct from 'biology' seems to have vanished" (Keller 1992:297). Keller suggests that this resurgence has been made possible because geneticists have succeeded in severing the connection between contemporary genetics and Nazi eugenics in popular conceptions of genetics today. Keller does not elaborate on whom she is talking about when using "we." Although her observations are quite compelling in regard to genetics in the United States, they are less so in the Netherlands.

For example, in the 1980s, as Keller notes, as biotechnology was taking off in the United States unchallenged, in the Netherlands, Schellekens and R. Visser, a historian of science, were writing a critical history of genetic practices. They argued that although there were eugenic movements around the world in the early decades of the twentieth century, there never was one in the Netherlands (Schellekens and Visser 1987).

In discussing his book with me, Schellekens explained that as a proponent of contemporary genetic research and practices, he felt it was important to understand the history of the discipline. He seemed relieved to be able to interpret the negative aspects of that history, that is, those associated with eugenics, as for the most part never having been taken up by Dutch people. The sense he communicated was that such practices were antithetical to Dutch sensibilities. Schellekens's and Visser's endeavor is not simply to demarcate the use from the misuse of scientific knowledge. In their book they argue that the Netherlands never had a prewar eugenics movement comparable to those of the United States, Great Britain, and Germany. Their work suggests that eugenic ideas were not suited to the Dutch character, then or now. Ann Stoler's (2002) analysis of Dutch eugenic ideas and practices in relation to its colonial efforts would seem to complicate the story Schellekens and Visser want to tell. In fact, there was a eugenics movement in the Netherlands in the first half of the twentieth century (Noordman 1989, 1994), when eugenics exploded as a worldwide phenomenon (Kevles 1995 [1985]; Paul 1995), and it seems to have been highly active in Dutch colonies. Nevertheless, Schellekens and Visser discursively work to diminish the significance and appeal of that movement in the Netherlands in comparison to its robustness elsewhere in the world.

Ambivalence about certain forms of difference has become significant in the Netherlands today. In his overview of Dutch society, William Shetter (1987) argued that the inability of a far right party to gain a strong foothold in the Netherlands at a time when such parties were thriving in much of Europe emphasized the lack of widespread support for such views in the polity. Shetter linked this phenomenon to the Dutch social ideal of tolerance. The rise of Pim Fortuyn and his anti-immigration party in 2002 as well as Fortuyn's assassination shocked the nation and the world because the two events challenged the ideal of Dutch tolerance. Anxieties about difference, of course, are embedded in contemporary genetic practices (Ginsburg and Rapp 2001; Taussig et al. 2003). I found that in the 1990s Dutch geneticists were working to emphasize the ways genetics could be practiced in the Netherlands without diminishing Dutch practices of tolerance.

A number of Dutch geneticists spoke with me about the need to develop a contemporary image of genetics, a need I interpreted as distinguishing a view of contemporary genetics free of its negative history or potential misuses. Two clinicians in their early thirties spoke directly to this point in discussing popular attitudes toward genetics with me. One of them, Geert, linked his discussion of popular attitudes toward genetics in relation to the Nazi legacy directly to the issue of Stier Herman:

> **KST:** How do you find people in the Netherlands reacting to advances in genetics?
>
> **Geert:** I think if it has medical implications I think most people support most things like that. You always do have quite a religious group that does not want to hear about it, but if it is a serious disease I think that most people support it and can imagine that people want to make use of the possibilities that are available in a center like this. . . . But you have probably heard about the Stier Herman?
>
> **KST:** I've heard about him, but I haven't followed too much of it.
>
> **Geert:** Yeah, yeah, well, there is a lot of debate going on. They have built a certain gene in a bull called Herman and the gene produces lactoferrine —that is a kind of enzyme that protects cows from inflammation of the udders. A lot of discussion is going on about that, about whether it is useful or not. I think Dutch people are quite . . . let's say . . . practical, if things do have use, make sense, then they agree to it, but if it is just to make the tomatoes bigger, well, then they're quite reluctant about it. And if it is to make the human race better than it was, well, then they are reluctant as well because, well, World War II of course did have a great impact on Holland, and the things that happened in Germany are constantly being remembered here in Holland. I think that has to do with [the way people interpret genetics]. So people are very aware, I think, of genetics and the things that go along with it, but if it has to do with disease I think they can agree to it.

Bart, the other clinician, also discussed his impressions of popular attitudes toward genetics. While he explicitly described popular identifications between genetics and Nazi science, he dismissed them as impractical:

> **KST:** How do you think Dutch people react to new possibilities that come from increasing scientific knowledge of genetics?
>
> **Bart:** It is really different from one person to another. . . . I cannot talk for science as a whole but I know that for genetics, what I do [clinical genetics], and I am not talking about patients but about the general population, the people that I know, they have a very negative perception of it. They think it is to enrich the species or, well, I think what happened in the past in Nazi Germany still is in the subconscious. It is

> lingering . . . in [attitudes toward] genetics so people are always thinking that genetics is a way to create a superspecies or something like that, which is conceivable, but on a practical level that is not at all what we are doing. But people are reacting to that image that they have in their heads when they think about new developments.

Geert and Bart acknowledge the importance of the public's concerns about genetic practices but do not themselves share them. Their response to the popular association of genetic research with the historical phenomenon of Nazi science is to emphasize specifics of genetic practice at the clinics where they work and to argue that these differ from the practices employed through the Nazi program of racial hygiene. A concern for such distinctions informed clinicians' understandings of their daily work.

Clinicians repeatedly identified two critical distinctions that set their practices apart from those that could be associated with inappropriate goals such as those associated with Nazi eugenics. These distinctions revolve around notions of choice and designations of "medical indications." At each stage of the process of genetic testing and diagnosis, geneticists work to ensure that patients retain the power to choose whether to participate. The clinicians most frequently deal with the issue of choice by stressing that genetic testing is done on a voluntary basis and that neither the state nor the genetics centers coerce people to undergo genetic tests. They try to ensure individual choice at the clinic level by employing a strategy of nondirective counseling in communicating testing possibilities to patients. That is, they see themselves as providing expert knowledge on the basis of which individuals can make informed decisions about reproduction.[19]

A number of informants, however, spoke to the way information available through genetic testing may constrain choice. Responding to my queries about the meaning of genetic conditions for society in general, virtually every person I spoke with raised the issue of the financial costs of caring for such people. In a society with a highly elaborated system of social welfare, the choice to knowingly have a child who is likely to put a strain on public resources often is considered a sign of irresponsible parenting.

Liesbeth, a twenty-nine-year-old taxi driver, went so far as to suggest that knowledge derived from genetic tests can coerce particular choices. She brought up the issue of borders in discussing prenatal testing, by questioning the kind of knowledge one gains through prenatal testing and the kinds of choices that knowledge forces. She explained that there had been a child with Down syndrome in her family, so when she was pregnant she had amniocentesis for prenatal diagnosis. "I am not sure I would want to do it again," she explained, "[a child with Down syndrome] cannot ever live alone . . . on the one side it [prenatal testing and abortion] may be against your principles but on the other side, if you know that it [having the child] will be truly very difficult . . ." Leaving this idea unfinished, she shifted to talking about knowledge. She said that today people know more and that they have "more and more questions and want to know what the [testing] possibilities are. The border [*grens*] is difficult. Where does the border lie?" In contrast to the geneticists' understanding of genetic knowledge as enabling free choice, Liesbeth experienced such knowledge as coercing choice.

The Dierenbescherming posters question the power of isolated individual choices to address these issues adequately. While the clinicians focus on choice exercised at the individual level in the encounter between patient and doctor, the Dierenbescherming insists on situating choice in a broader context. They suggest that choices about biotechnology can be meaningful only if located through politics at the societal level.

In addition to locating choice at the individual level, medical geneticists in the Netherlands emphasize that genetic counseling and all subsequent testing are based on a recognized medical indication, and the guidelines for eligibility for genetic testing in the Netherlands have been established around this concept. These guidelines also include a policy of not conducting genetic tests on minors, unless such tests are considered essential to the child's health. Geneticists believe that their reliance on recognized medical indications for conducting tests distinguishes their practices from those associated with the Nazis and other eugenic practices.[20] In the case of someone seeking prenatal testing, such indications include an increased likelihood of giving birth to a child with a severe physical or mental disability. During my fieldwork these indications typically included maternal age, family history, and an individual's own prior reproductive history. Questions raised about whether there was a medical indication for testing were brought up in clinic meetings explicitly in terms of the need to resolve ethical questions. Sometimes such questions were raised on a theoretical level and sometimes a clinician had a case in which the medical value of testing was ambiguous. In both instances the goal was to establish consensus about what constituted a medical indication for genetic testing.

Defining what constitutes a medical indication, however, can be more problematic than the comments made by Geert and Bart might indicate. For example, one day at one of the clinics where I conducted fieldwork, the physicians began discussing whether or not a family history of polyposis coli (a form of colon cancer) could be an indication for prenatal diagnosis. The physicians explained to me that in many cases genetic testing for a strong predisposition to this disorder is possible. According to the genetics literature, if one has the dominantly inherited gene in question there is "nearly a 100 percent chance" of getting colon cancer, which ultimately requires a colostomy. The problem, they emphasized, is that polyposis coli carries no increased risk of mental or physical disability as defined by the clinicians. According to the clinicians, the couple seeking the prenatal testing were quite insistent that they did not want to be responsible for passing polyposis coli on to their children. When the clinic staff initially resisted their request, the couple insisted that their personal experience with the disorder enabled them to determine that polyposis coli was a severe handicap. One clinician explained their response, telling me they had said, "You might think there is not a medical indication for this test, but we have lived with it and we *know* what it is like, so you will *never* convince us that it isn't a serious medical problem." In discussing this case, another clinician observed that it is "difficult to say [to someone who has this disorder], 'You have to accept this risk [a 50 percent chance of having a child with the gene]' because it isn't your life, and you don't have to raise this child."

Nevertheless, once the couple had convinced the clinicians there might be a medical indication for the test, the obstetricians at the teaching hospital, that is, the practitioners who actually do the amniocenteses that provide the cells for prenatal tests, resisted. They did not agree with a designation of a medical indication for polyposis coli as acceptable for prenatal testing. One geneticist discussed the situation vis-à-vis the obstetricians with me, saying they had a right to have a say in the matter "because, after all, they are the ones who do the tests and they are the ones who have to do the abortions when there are bad results." After a series of conversations between the geneticists and the obstetricians a consensus was reached that the experience of the couple in question should be taken seriously. In the end, they got the test. The couple succeeded because they were able to invoke their personal experience as a means to contest and negotiate the meaning of what constituted a medical indication. The patients' success reflects both a concern for the agency of the patient and sensitivity to the image of coercive Nazi eugenics among Dutch medical professionals.

Whether the clinicians actually are ensuring individual choice or maintaining the medical basis of genetic services, the ideological work they are engaged in involves distinguishing their practices from those associated with the powerful symbol of Nazi eugenics. To accept these distinctions, Dutch people must trust in the professional integrity of genetic practitioners to believe their assertions about choice and medical necessity. The Dierenbescherming posters graphically call into question the relevance of such distinctions by situating genetic knowledge and practices in a larger historical, political, and cultural context.

GENETIC POLITICS AND CULTURE

Many societies are well attuned to the history of Nazi eugenics. Many societies have cows. Nonetheless, the ways the Dierenbescherming staff, geneticists, and others in the Netherlands connect ideas about Nazi eugenics, government regulation, modernity, immigration, cows, national identity, the nation, milk, purity, race, gender, and contemporary biotechnoscience are distinctively Dutch. The continuing desire of many Dutch people to define themselves against their more powerful German neighbors, living memories of the Second World War, anxieties about increasingly porous borders and new immigrants, and the Dutch social ideal of tolerance combine powerfully to shape genetic knowledge and practice across a wide array of social domains in the Netherlands. Geneticists, couples seeking genetic counseling, political activists, members of parliament, documentary filmmakers, ordinary citizens, and multinational corporations have all engaged with these local issues in their own way.

In this chapter I have explored the distinctive cultural repertoire Dutch people draw upon to make sense of contemporary biotechnoscientific knowledge and practices related to genetics. Despite geneticists' claims to authoritative knowledge, my analysis indicates that genetic knowledge and practices are not simply biology and not the exclusive domain of scientific, technological, and medical practices. Other groups and individuals continually engage in constructing competing and sometimes conflicting interpretations of the significance of genetics in contemporary Dutch society. Although using the services available at Dutch genetics centers may serve as a marker of a positive conception of a modern Dutch identity, I also have shown that through its association with popular understandings of the legacy of Nazi science and the potential transgression of socially valued categories, genetics also may serve as a powerful negative symbol of contemporary life. My illustration of encounters among contemporary understandings of genetics, gender, anxieties about recent social change, the social value of cows, race, the historic legacy of the Second World War, and the significance of tolerance in light of the Dierenbescherming's posters makes visible the cultural and historical specificity of constructions of genetics in the Netherlands. It also highlights contemporary challenges to the Dutch social ideal of tolerance and thus to nation and identity in the context of recent social change and persistent ambivalence about certain forms of difference.

The Dierenbescherming's challenge to biotechnology capitalized upon the potent symbolism of cows in Dutch society and drew directly on widespread attitudes toward the Second World War and Nazi science. The force of the posters derived additionally from the way these powerfully negative symbols played against the Dutch social ideal of tolerance and the strain new immigration has put on that ideal in the Netherlands today. In drawing on broadly held concerns about genetics, borders, the nation, and the body to publicly question and protest genetic practices in the Netherlands, the Dierenbescherming indirectly raised the problems many Dutch people associate with practices in human genetics. At the same time they portrayed such problematic uses of genetics as antithetical to Dutch identity.

The persistent salience of the legacy of Nazi science in the Netherlands and the centrality of tolerance to conceptions of Dutchness allows groups such as the Dierenbescherming to invoke history as a way of contesting the meaning of genetics in contemporary Dutch society. But invoking the Dutch ideal of tolerance also calls attention to contemporary concerns about whether and how Dutch people can rely on this ideal in light of rapid social change. Dutch geneticists are continually engaged in the ideological work of defining their practices as distinct from those of the Nazis. Nevertheless, I have shown how the very principles of individual choice and medical integrity through which geneticists forge such distinctions remain open to competing interpretations by both individual patients and such groups as the Dierenbescherming. Elaboration of these processes demonstrates how genetics in the Netherlands has given rise to negotiations and contestations over the meaning of genetics, the body, the person, and the nation within and across a wide range of social domains. The locally distinctive nature of these processes illuminates the enculturation of the production, consumption, and practice of genetic knowledge in the Netherlands today.

Epilogue

ORDINARY GENOMES IN A GLOBALIZING WORLD

We can investigate globalist projects and dreams without assuming that they remake the world just as they want.

—Anna Tsing, "The Global Situation"

The life sciences constitute an area of knowledge that works in fast forward. Nearly every day there are media reports about new discoveries in the rapidly expanding field of genetics and genomics. These reports detail the latest news of genetic explanations for a wide range of human afflictions, from breast and other cancers to schizophrenia, autism, and Huntington's disease. They also describe suggested links between genetics and behavioral or personality traits; the gene "for" homosexuality, the gene "for" shyness. Yesterday's science fiction fantasies involving phenomena such as cloning and other new reproductive technologies have entered today's reality at an increasingly rapid rate. Unfulfilled claims about the possibility of gene therapies have been replaced by promises that stem cells may be turned into regenerative treatments that will improve health and extend life.

Ultimately, in those countries where it is socially relevant and culturally available, genetic knowledge and practices affect the way we humans think about ourselves and what it means to be human. These issues, however, do not exist in a vacuum. If the human genome is accepted as a Holy Grail, if we are conceived of as the sum of the base pairs making up our DNA, then we risk marginalizing history and social practice as significant forces in our lives. We generally understand that developments in genetic knowledge and practice influence society. This common view assumes that the biological is somehow prior to the cultural. What is less fully appreciated is how our views of the world shape both the way we think about genes and the ways scientific and medical knowledge of genetics is produced and taken up in everyday life. In other words, as I have argued in this book, genetics may transform society, but society also transforms genetics. What, precisely, genetics is and will be is being worked out in the kinds of daily practices I have described. To fully understand the significance of genetics one must appreciate how genetic knowledge and practices are distinctively generated and reproduced in specific local contexts and across diverse social domains.

This book is a study of the global enterprise of genomics in a local setting. Analysis of the Dutch case contributes to our understanding of how scientific knowledge is always and ever entwined with culture. Many of the issues and practices explored in this book prefigure more recent, and ongoing developments in the life sciences. Indeed, as I have aimed to show, learning to see the life sciences as a multiply constituted endeavor implicating global and local, normal and abnormal, citizen and nation will be an essential tool for understanding technoscience as it continues to unfold. Looking forward from this study toward ongoing developments in the life sciences, I see these issues figuring prominently in contemporary efforts in biobanking and in stem cell research. Each is a globalized phenomenon that nevertheless plays out in specific national contexts with particular cultural commitments.

BIOBANKING

Since the completion of the Human Genome Project, the development of nationally identified biobanks has emerged as a central component of ongoing efforts to capitalize on new biotechnologies.[1] National biobanks typically involve the storage of DNA samples from as many citizens as possible combined with a system for linking that DNA to both genealogical records and medical histories. Researchers consider the combination of these three resources—DNA, medical records, and family histories—the essential components of translating new genetic knowledge into clinically useful medical interventions. Estonia, Iceland, Japan, Korea, Latvia, Singapore, Sweden, and the United Kingdom, among others, have all established national genetic biobanks (see, e.g., Fortun 2008; Lee 2006; Tutton 2007). These banks are explicitly intended to facilitate molecular research in these countries as a way of maintaining or achieving recognition for being a site for significant scientific innovation or otherwise contributing to the production of scientific knowledge. A number of these efforts also have embedded in them an idea of a distinctive national genome. The enterprise of creating a Korean national biobank is undergirded by an understanding that the "Korean Genome" is somehow "different from the human genome" (Lee 2006:443). Such a conception suggests that "the human genome" produced by U.S. researchers gains the status of an unmarked molecular norm. The link between nation and genome can also be seen in France, where a national scandal erupted when the Centre d'Étude du Polymorphisme Humain (CEPH), the French biobank, attempted to strike a deal with an American biotech company. The controversy stemmed from concern that France's genetic patrimony was being commercialized and sold to America. The French government ultimately prohibited the deal (Rabinow 1999).

As supposedly national genomes proliferate, they are becoming distinct, bounded, nationally identified genetic norms that exist alongside, yet separate from, a global norm of the human genome. The issues raised by national biobanking, including national identity, race, purity, modernity, and normalization, echo those we find in the Dutch case.[2] The case of Iceland and deCODE Genetics offers an instructive illustration of how these themes are playing out as people work to develop the social arrangements that facilitate contemporary genomic research.

In the late 1990s Kari Steffansson, an Icelandic physician and researcher working at the University of Chicago, convinced the Icelandic government to create a database of Icelanders' medical records that would be combined with a database containing genetic material (DNA) from Icelanders that Steffanson would create and bring together with Icelandic family genealogies. The aim was to pursue the kind of genetic research now coming to be recognized as essential to developing molecular interventions into human health.[3] In exchange for the agreement of the government to facilitate access to these material means of contemporary knowledge production, Steffansson returned to Iceland and established the biotechnology company deCODE Genetics, which was supported almost entirely by the Switzerland-based multinational pharmaceutical firm Hoffmann-La Roche. The explicit aim of deCODE is to mine these resources for clinically useful findings (Fortun 2001, 2008; Rose 2001; Sigurdsson 2001). The establishment of deCODE Genetics turned virtually every Icelandic citizen into a research subject (Fortun 2001, 2008; Rose 2001; Sigurdsson 2001; Specter 1999).

The willingness of the Icelandic government to create and grant exclusive license for a health records database to a single commercial enterprise generated controversy, although, according to polls, the majority of the Icelandic population supported the effort (Sigurdsson 2001). The complicated and fractious process of obtaining consent from the people of Iceland to use their genetic, medical, and genealogical data for the national biobank also implicated discourses of genetics as a site for the articulation and fulfillment (or contestation) of the duties of modern citizenship (Taussig 2005).

In working to convince Icelanders to commercialize their national genome, Steffansson linked genes to nation and geography to articulate and promote the project in terms of pride in a distinctive Icelandic historical and biological identity. Steffansson argued that because of its geographic isolation and the population's purported genetic homogeneity, Iceland was uniquely well suited to the kind of genetic research he was proposing. Claims about genetic homogeneity, which are highly contested by population geneticists (Arnason 2000; Lewontin 1999), speak forcefully to widespread ideas about the purity of Icelandic identity that reckon Icelanders as direct descendants of the original Vikings who founded the nation. It is, in fact, just this commitment to the past that fuels Icelanders' apparent passion for genealogy, a phenomenon that Steffansson explicitly recognized as providing an important component of facilitating genetic research. Further, Steffansson's effort spoke to concerns about the Icelandic nation's ability to play a meaningful role in the global economy. In the coverage of the establishment of deCODE one finds articulations of anxiety about brain drain involving the departure of Icelandic professionals and the nation's failure to lure those, such as Steffansson himself, who had obtained higher education and graduate degrees in foreign lands to return to Iceland.

These events occurred simultaneously with an increased recognition of the decline of the ocean's fish stock, something of much concern in a nation where fishing had long been the primary industry. In arguing for the creation of and access to a national biobank, Steffansson agreed not only to return to the country, but also to establish a new industry in Iceland and to create jobs for other Icelandic professionals. In so doing, he articulated a vision of a modern nation that did not just supply the world with fish and fish products but also made contributions to cutting-edge scientific knowledge. In supporting this effort, the nation moved to both claim and project a modern identity based on a commitment to contemporary endeavors in the life sciences. It is thus hardly surprising that by the summer of 2000 at least some Icelanders were taking pride in the idea that their genetic material was being used to further scientific research and the national tourism board was using images invoking science labs to depict the country (Tringe 2001).[4]

Like the idea of a genetic passport and efforts by the Dutch to make sense of genetic difference, national biobanks reference national identity, race, purity, modernity, and normalization. By creating—mapping—distinct national genomes those involved are working to create their own prototypical, nationally identified genome. In seeking to exploit such resources for the development of new knowledge and medical interventions, these nations articulate a modern identity. In these processes two distinct dynamics are at work. First, genomes are becoming sites for asserting national identity; second, nations are using their national identity to produce prototypical genomes and claims about their status in relation to modernity. Pervading these dynamics are market-driven concerns literally to capitalize upon national genomic resources in a globalizing biotechnology marketplace. In this context, entering the global arena of biotechnology markets is, ironically, conditioned upon assertions of distinctly local national genomic identities. Scientific prestige and national economic power are increasingly bound together in producing genomic futures (Fortun 2008) at both the individual and the national level. In all of this one sees, as in the Dutch case, the invocation of local history and geography to connect genetics and national identity while simultaneously promoting a nationally specific strategy for the exploitation of emerging biotechnologies.

STEM CELL RESEARCH

Embryo research, cloning, and the creation of genetic chimeras are at the heart of scientific practices associated with contemporary stem cell research. Stem cell research gains its public visibility through the dramatic promises scientists make about the potential of stem cell therapies to intervene in human health and life. In the United States stem cell science has also become public through the ways it has extended particularly American reproductive controversies having to do with abortion politics. The idea of stem cell therapy is grounded in the potential to manipulate stem cells to develop into distinct, useful cells (e.g., a skin cell, a liver cell, or a blood cell). The cells that researchers believe have the greatest potential to become any kind of cell are embryonic stem cells. With current technology, gaining access to embryonic stem cells involves destroying embryos. In the United States the destruction of embryos for stem cell research as part of the effort to investigate the potential of stem cell therapies thus places this research enterprise at the center of abortion politics.

In this context, again, biotechnology is invariably locally embedded. The meaning of and practices regarding an embryo vary depending upon national context. That an embryo might have distinctive meanings was made particularly clear to me when, during my fieldwork in the Netherlands, researchers at a conference in Montreal, Canada, claimed to have cloned a human embryo. This claim garnered headlines around the world (e.g., Kolata 1993). As it turned out, what was announced as cloning in this case from 1993 is quite different from what is today described as cloning or somatic cell nuclear transfer—the process that produced Dolly the sheep in 1997.[5] In fact, at the conference in Montreal the physician Jerry L. Hall of George Washington University, a fertility specialist, announced that he had divided seventeen embryos into forty-eight embryos, describing this as cloning.

When the news of this embryo cloning broke in the Netherlands, the Dutch geneticists with whom I was working expressed utter disbelief that an embryo could be divided into clones. When they investigated the reports further, they realized that what had been cloned (or, actually, divided) was not what they considered an embryo, but an entity known in the Netherlands as a pre-embryo, that is, an embryo in the first fourteen days of existence, before its cells have begun to differentiate. The Dutch researchers regarded an embryo and a pre-embryo as distinct entities. Later, I was interested to hear American researchers who learned about the distinction made in some countries between an embryo and a pre-embryo suggest that such terminology might help them navigate the complicated politics of embryo research in the United States. In the United States, the status of the embryo is caught up in debates about when life begins and the boundaries between human and nonhuman. Perhaps it is not surprising that in the context of current controversies over stem cell research in the United States the word *blastocyst* has become increasingly widely used to talk about this biological entity. The meaning and significance of embryos are unstable, differing across national contexts. These differences matter. They also provide an example of how the biological and the cultural inevitably are intertwined, together producing the entity known as an embryo.

The problem of the use of embryos in stem cell research in the United States has, under the administration of George W. Bush, led to restrictions in federal funding of such work. In fact, in the wake of 9/11, it is often forgotten that President Bush's first major policy pronouncement came in August 2001, when he set forth limits to federal funding for stem cell research. Scientists in the United States reacted by asserting that such restrictions will curtail stem cell research in the United States, with the potential result that scientific establishments in Asia and Europe could surpass the United States as sites of important scientific and commercially valuable innovation in this area. Reactions to the policy have also raised the specter of a scientific brain drain characterized by an exodus of American scientists moving to countries where different ideas about embryos mean that gaining access to the materials necessary for pursuing stem cell research is not so restricted.[6] Such reactions thus explicitly link scientific accomplishment to national identity and to remaining at the forefront of efforts in biotechnology that are themselves connected to understandings of modernity. Moreover, in the religiously charged context of abortion politics, many advocates of stem cell research also contrast the modernity of biotechnology against the purported backwardness of Bush's religiously inflected policies.

One technique stem cell researchers rely upon to investigate the potential of stem cells as therapies involves the creation of human-animal chimeras. In many respects the Dierenbescherming's poster of the woman with cow udder breasts, circulated in response to the transgenic bull Stier Herman, was playing on popular discomfort with boundary-crossing chimeras. Stem cell researchers create chimeras that similarly embody genetic material from both humans and other animal species.[7] Ethical issues about human experimentation underlie contemporary stem cell researchers' development of chimeras as a tool for stem cell research. Many researchers argue that the ability of human stem cells to proliferate and function as well as their adverse potential to become tumors can best be studied in embryonic environments. Concerns about human experimentation prevent them from conducting such studies in human embryos, so they see the nonhuman embryos of human-animal chimeras as an appropriate tool for investigating these issues (Robert and Baylis 2003). In such cases the laboratory models become assays for investigating the safety and the potential of stem cell therapies. Some researchers are concerned about the use of chimeric organisms in stem cell research. As Robert and Baylis point out, these researchers "are quite sensitive to the ethical conundrum posed by the creation of certain novel beings from human cellular material and their reaction to such research tends to be ethically and emotionally charged" (2003:2). One of the intriguing aspects of this avenue of research has to do with the practical problem of determining whether and, if so, precisely what animal(s) would constitute an appropriate model for investigating the efficacy and safety of stem cells in humans. Although at a different scale than in the case of Stier Herman, these are questions about the identity between humans and animals, and they reveal widespread concern about the use of genetics to cross or blur species boundaries.

As debates over the naming and nature of embryos implicate understandings of the boundary between human life and not life, technologies creating chimeras involve uneasiness over boundaries between human and not human. Robert and Baylis suggest that anxieties about the hybrid nature of human-animal chimeras may also stem from the ambiguous moral status of such organisms and their potential implications for existing and future relationships.[8] In some respects, the boundaries between human and nonhuman present more broadly troubling issues, issues that worry even those who embrace stem cell science. Here again one sees issues foreshadowed by the Dutch case. The case of Stier Herman also involved the biotechnological production of an interspecies entity. Dutch interpretations of Herman as the limit case for the appropriate use of biotechnology suggest that as scientific practices continue to test the limits of public acceptance of the crossing of socially valued boundaries between human and nonhuman, one can expect to see diverse, locally articulated, and socially informed responses to such new technologies.

Stem cell research is also a prominent site of the articulation of national identity and prestige through the public and political embrace of scientific practice. Countries like Singapore, England, and Korea actively promote stem cell research. Singapore's efforts in this regard resonate with American researchers' anxieties about brain drain in that these efforts explicitly seek to build the capacity for human capital needed to gain world recognition as a site in which scientists are conducting world-class cutting-edge research (see, e.g., Arnold 2006; Ong 2005). Such efforts are frequently linked to the nation and to national honor. In Korea, when Hwang Woo Suk and his colleagues announced the first successful cloning of human embryos he was lionized as a national hero and had numerous national honors bestowed upon him, including having his likeness placed on a postage stamp. Hwang's claims were later disproved in what devolved into a notorious case of scientific fraud involving major international peer-reviewed scientific journals. The prominence accorded the fraud only serves to underline the strength of the bond between scientific accomplishment and national prestige exemplified by the response to Hwang's claims.

Furthermore, in efforts to establish scientific credibility with regard to stem cell research, national identity and economic advantage are linked in the notion of scientific advancement. Biotechnology currently represents the acme of modern scientific and technological achievement. In the cases of Korea and Singapore (as in the Iceland example of biobanking) the scientific achievements signaled by success in cutting-edge scientific research also serve to make a claim to a particular understanding of modernity involving scientific rationality. Kaushik Sunder Rajan identifies a similar phenomenon in the practices of Indian scientists working in biotechnology who link salvation and nation in "the role they see for themselves as agents in India's development" (Rajan 2006:35). Such efforts resonate with the dynamic through which modern Dutch identity is being crafted in and through contemporary genomics.

ORDINARY GENOMES

Scientists, and bioethicists who focus their attention on individuals, typically do not consider the effect of emerging knowledge and technologies on national cultures and global politics more generally. Yet, as new knowledge arising from human genome research and its application in medical practices reshapes people's understandings of nature, individual and collective views of the world and of ourselves also shift. The proliferation and dissemination of genetic knowledge increasingly force one to think of health, the body, ability, identity, and reproduction specifically in terms of genetics. Such concepts as a genetic passport and descriptions of the human genome as a Holy Grail point to how profoundly genetic knowledge and practices currently are reshaping the public's understandings of what it means to be in the world.

Genetics is thus becoming an increasingly powerful system for defining what counts as normal, healthy, acceptable, and even human. In June 2007, David Brooks, a columnist for the *New York Times*, editorialized about a Harris poll showing that 40 percent of Americans "would use genetic engineering to upgrade their children mentally and physically." He went on to state that "if you get social acceptance at that level, then everybody has to do it or their kids will be left behind." Beyond the obvious resonances of eugenics that such phenomena raise, they also bear on definitions of normality that implicate a broad array of political, social, and personal questions related to reproduction, health, access to health care, insurance, and identity. But, as the Dutch case illustrates, assumptions about what counts as normal are shaped in locally specific and sometimes unanticipated ways. Genetics is a rapidly developing field, one that is still in its infancy. How it will develop and what ultimate possibilities it will offer is unknown, but we do know that diverse societies will continue to struggle with the meanings and implications of genetic knowledge and its applications in daily practice.

I have explored how in the Netherlands genetic knowledge and practices are caught up in Dutch fears and fantasies about what it is to be human as well as Dutch, normal, and a citizen. In looking at the historical and cultural specificity of scientific knowledge and practice in a particular place at a particular time, I have argued that scientific knowledge and practice, as the products of lived experience, are powerfully inflected by local contexts. In analyzing the case of genetics in the Netherlands, one sees that knowledge about genetics constitutes and is constituted by the distinct social context of the Netherlands and, at the same time, that what the Dutch context itself means in the age of the European Union and of international genomic medicine is itself in transformation. Ordinary genomes are a distinctively Dutch phenomenon, but they are also deeply imbricated with globalizing biomedical technologies. Neither exists without the other, but neither subsumes the other. They have been and continue to be produced together in social practices that are at once collaborative and contested.

The concept of ordinary has been at the center of my analysis. For many in the Netherlands, at some level, to be ordinary is to be Dutch. At the same time, I have aimed to show that this is a contested term with many meanings for different players within Dutch society. How the ordinary citizen will be imagined in the genetic age takes shape within the frame of Dutch commitments to tolerance and equality—defined as much by their centers as by their limits—within ideas about the secular and the religious in the Dutch polity (especially in relation to the imagined place of Orthodox Reformed Calvinists) and against the backdrop of fears rooted in the Second World War that genetics will again step into the territory of eugenics. In the Dutch case, Foucault's concept of biopolitics sheds light on a nationalist imagination that holds normalcy and ordinariness not only as regulatory ideals in Foucault's sense, but also as explicit terms that people in the Netherlands use every day to articulate cultural values.

My book also complicates any simple identification of globalizing genetic science with the social category of the West. Dutch people, including Dutch geneticists, work hard to render genomes ordinary, but this confounds claims that genomes are natural. As genetics becomes more ordinary, in the routine sense, in different ways in different settings around the world, it is critical to remember the intricate tangle of forces—local and global, biophysical and cultural—that combine to produce what we call genetics and the biological writ large. The complex, distinctive ways genetic knowledge is produced, circulated, and taken up in daily practice in a quintessentially Western country such as the Netherlands thus dissolves the notion that there is any such thing as a monolith of Western modernity or, indeed, of modern bioscience. However the life sciences and the possibility of new interventions in human health develop, their meanings and practices will unfold differently in different contexts. Recontextualizing the otherwise often vigorously decontextualized accounts of the technical problem of the progression of laboratory results in clinical medicine is an essential component for understanding how new biotechnologies are adopted and negotiated inside culturally defined sets of local meanings. As we continue to struggle to understand the meaning and significance of emerging biotechnologies, it is important to keep in mind the many ways they are constituted in and through the interplay of cultural effects—including local, national, and global aspects of power, aesthetics, geography, and memory—that transform and complicate what are usually glossed as universal scientific objects and events.

NOTES

INTRODUCTION

1. All stand-alone first names as well as the names of small villages in this book are pseudonyms. In some cases I have changed specific identifiers to protect the privacy of those who were generous enough to share their time with me by participating in the research on which this book is based.
2. As a mapping of Dutch colonialism, her comment offered a telling reminder of the history of anthropology and colonialism (Dirks 1992).
3. A rough draft of a map of the human genome was announced in 2000, and the completion of two different versions of the draft were published in 2001. This work has been the foundation for a significant expansion in scientific knowledge about life. However, the dramatic promises for intervention into human health and life have largely been unrealized (Lewontin 2001; Taussig 2005).
4. For history and analysis of the passport as a document of identity and citizenship, see Caplan and Torpey (2001) and Torpey (2000).
5. Eugenics, which I discuss further in chapter 2, is the word used to describe projects that aim to improve the human species through the encouraging and discouraging of specific human matings. Its most obvious elaboration was in the Nazi project of producing an allegedly pure Aryan race by promoting reproduction among those designated as Aryan and preventing the reproduction of others, most visibly through the effort to exterminate Jews, homosexuals, Gypsies, the disabled, etc.
6. Analysts of Dutch society argue that the historical necessity of constantly working together to drain water from the land forged a unique national identity that promotes social interaction and cooperation (Goudsblom 1967:140; Schama 1988:15–50). That other analyses of the Netherlands almost invariably touch on the presence of water in Dutch cities and in the consciousness of Dutch citizens (Shetter 1987; Stephenson 1989, 1990) demonstrates the significance of these themes in the construction of identity in the Netherlands.
7. Generally speaking, anthropologists have considered sites remote from the cosmopolitan urban centers that are home to most of them and their institutions as the traditional and appropriate place for anthropological inquiry. A

nuanced critique of this stance is now extraordinarily well rehearsed in anthropology but, as Gupta and Ferguson (1997) point out, a hierarchical ranking of the sites of anthropological fieldwork persists in the discipline. The more remote a site is from the anthropologist's social and geographical location (predominantly the urban, industrialized nation-states of Europe and North America, where academic institutions continue to train most of the world's anthropologists), the more status she attains. During the period of decolonization in the decades following the Second World War, access to anthropology's preferred field sites became more difficult. As a result, John Cole suggests, many anthropologists turned their attention to places like Europe (Cole 1977:355–58). Even as they did so, however, anthropologists still tended to seek out field sites on the periphery of urban industrialized Europe (Arensberg 1968; Barnes 1954; Blok 1975; Chapman 1971; Schneider and Schneider 1976). Although geographically located within the West, such sites were perceived as being socially and technologically premodern and therefore remote in time if not in space.

8. This is visible in science studies texts that focus on the laboratory or the clinic. Ironically, such texts often illuminate the subtleties of local practices within the laboratory or the clinic while completely ignoring the relationship between such sites and their local context outside the laboratory. Consider, for example, Emily Martin's description of being admonished by Karin Knorr-Cetina for not staying put in the laboratory (Martin 1994, 1997).
9. In the era of globalization it is useful to keep in mind that science, with its claims to universality, has been, from its beginnings, a powerful world-making project. European science flourished in the eighteenth century not merely because of the intense curiosity of a few individuals in Europe, but also because it extended the Enlightenment project of discovering, knowing, and dominating the world (Beer 2000; Dirks 1992). In this sense science is intimately tied to other dominant strains of European history having to do with exploration and colonization, a history in which the Netherlands and the Dutch East Indies Company played significant, highly specific roles. Anthropology, cartography, geography, botany, and all of the natural sciences were essential components of an imperial expansion that both necessitated and facilitated the active exercise of a scientific imagination organized around knowing and controlling nature (Dirks 1992).
10. Indeed, as Donna Haraway cautions, demonstrating the historical contingency of "scientific and technological constructions" risks becoming a project of sorting good from bad science rather than one of finding what she describes as more real accounts of the world (Haraway 1991:187). For Haraway, developing these richer accounts of the world requires a clear understanding of how meanings and bodies are produced in and through the lived experience of scientific

knowledge and practice. It is in these processes that one can view a more complete, nuanced understanding of how science is part and parcel of everyday life.

11. Genetic practice is thus a site for exploring how local forces shape both the production and the consumption of global practices. In her work on globalization Anna Tsing argues that globalization needs to be understood as a set of projects. She points to the need to trouble our conceptions of "global forces" as practices that transcend place and the local as the place where "global flows fragment and are transformed into something place bound and particular" (Tsing 2000:338). She encourages anthropologists to elaborate "the cultural specification of the cosmopolitan" because "there can be no territorial distinctions between the 'global' transcending of place and the 'local' making of places" (Tsing 2000:338).
12. The industry's potential for producing both cultural and actual capital has meant that biotechnology has been widely identified by numerous national governments, including the Netherlands and the European Union, as an industry that is highly attractive to recruit. Iceland (Fortun 2001, 2008), India (Rajan 2007), Singapore (Ong 2006), the United Kingdom (Franklin 2007; Tutton 2007), and France (Rabinow 1999) are just some of the countries that have gained both academic and media attention for their efforts in promoting biotechnology.
13. Since the funding of the Human Genome Project numerous books have been published on genetics, the genome project, and the implications of new genetic knowledge. Often intended for a broad audience, these publications, written by journalists, bioethicists, lawyers, biologists, social scientists, and historians (e.g., Andrews 1999; Beurton et al. 2000; Cook-Deegan 1994; Hubbard and Wald 1993; Kay 2000; Keller 2000, 2002; Kevles and Hood 1992; Lewontin 1991, 2000, 2001; Lindee 2005; Nelkin and Lindee 1996; Ridley 1999; Rothman 1999; Suzuki and Knudtson 1990; Goodman et al. 2003), offer histories, analyses, and critiques of new genetic knowledge and its implications.
14. The literature on the history of science and scientists in the Netherlands, even in English language publications, is voluminous. Some primary themes in this literature involve strong connections between religious structures and scientific practice, an emphasis on practical purposes of knowledge, a focus on the presentation of scientific knowledge to the general public, and an effectively centralized structure for the dissemination of this knowledge. See, for example, Beek (1985), Bell (1950), van Berkel et al. (1999), Ford (1991), Huerta (2003), Roberts (1999), Simonutti (1999), Yoder (1988).
15. In her analysis of the concept of the "normal genome" in twentieth-century evolutionary thought, the philosopher Lisa Gannett points out that geneticists with more evolutionary perspectives, such as genetic anthropologists and popu-

lation geneticists, have criticized the molecular genetic Human Genome Project for "treating genetic variation as deviation from a norm" (Gannett 2003:146) and for its "constitution of a standard of genetic normality" (Gannett 2003:182). She goes on to illustrate that while the reliance on a concept of a normal genome may be most obvious in molecular genetics, the concept is deeply embedded in twentieth-century evolutionary thought.

16. Further, because the normal is defined only in relation to the abnormal, the abnormal is always prior to the normal. Canguilhem argues that "the abnormal, as ab-normal, comes after the definition of the normal, it is its logical negation. However, it is the historical anteriority of the future abnormal which gives rise to a normative intention. The normal is the effect obtained by the execution of the normative project, it is the norm exhibited in the fact. In the relationship of the fact there is then a relationship of exclusion between the normal and the abnormal. But this negation is subordinated to the operation of negation, to the correction summoned up by the abnormality. Consequently it is not paradoxical to say that the abnormal, while logically second, is existentially first" (1991:243).
17. Such ambiguity, Keller tells us, leaves open interesting possibilities for human agency in "how the authority for prescribing the meaning of 'normal' is distributed" (Keller 1992:299). Keller argues that in this phenomenon culture, "hidden from view" exerts its "undeniable force" (Keller 1992:299).
18. This is what Kim Fortun describes as a "place where the future is worked out now" (2001:361). Fortun points out that this is the space of the *anterior future,* elaborated by Derrida. Fortun helpfully elaborates this phenomenon, pointing out that the future is anteriorized when the past is folded into the way reality presents itself, setting up both the structures and the obligations of the future. "The future anterior is not unlike determinism, but dispersed rather than operationalized through linear causality. The future inhabits the present, yet also has not yet come" (Fortun 2001:354). I thank Kim Fortun for calling my attention to the significance of this space in regard to contemporary genetics.
19. The compression of the experience of time and space and related pressures of globalization mean that today people experience a world marked by the flow of people and things that challenges prior perceptions of the stability of the nation and national identity. The literature on citizenship and on globalization is large and growing. For review and analysis of major themes on citizenship, see Ong (1995, 1996, 1999), Shafer (1998), and Torres, Inda, and Mirón (1999). For work on globalization see Lechner and Boli (2004), Inda and Rosaldo (2001), and Tsing (2005). Much of this work deals with the intersection of questions of citizenship and those related to globalization.
20. A number of researchers working to understand the production and meaning

of new scientific knowledge have encountered the issue of citizenship, proposing concepts such as "somatic citizens" (Novas and Rose 2000), "corporeal citizens" (Lock 2004), "biological citizenship" (Petryna 2002; Rose and Novas 2005), "biopolitical citizenship" (Epstein 2007), and "genetic citizenship" (Heath, Rapp, and Taussig 2004). In one way or another, each of these authors is working to come to grips with the reconfiguration of the relationships among science, subjectivity, and citizenship set in motion by global transformations and new knowledge in the life sciences.

21. In elaborating a related concept of biological citizenship Rose and Novas (2005) argue that with this new knowledge come new forms of personal responsibility in which the ability to know produces responsibility for knowing and for taking charge of one's genomic future.

CHAPTER ONE: "GOD MADE THE WORLD"

1. Dutch tolerance and its relationship to scientific inquiry was even noted in 1685 by John Locke (quoted in Schaffer 1989).
2. These additional characters include Danielle's lesbian lover and partner; Thérése, Danielle's daughter, the issue of a one-night sexual escapade planned explicitly for the purpose of conception—Thérése is extraordinary in her intellectual brilliance and her ability to connect with Finger, from whom she learns physics and philosophy while still a young child; Thérése's paternal aunt, a single mother with two children, who arrives with her children, unannounced, from the city, looking for a new home; the former local priest, who has left the church because he finds its demands too repressive; Sarah, Thérése's daughter, unusual in her intense observation of life and death and the border between the two.
3. For example, in his analysis of race in what he calls the "Dutch world," the historian Allison Blakely points to the pervasiveness of pluralism and understandings of tolerance in the Dutch context:

   One of the most interesting themes [in Dutch social life] is that of pluralism, often noted as one of the most characteristic features of Dutch society. Often paired with the concept of tolerance, this Dutch pluralism has contributed much to the reputation the Netherlands has enjoyed at home and abroad as a haven for religious and political refugees. While the themes of pluralism and toleration by no means encompass all aspects of Dutch culture, they have bearing on . . . many. . . . The essence of what is often referred to as tolerance in Dutch society is a consensus that disagreement and diversity need not mean disharmony. . . . This attitude has allowed the Netherlands since its beginnings in the sixteenth century to reconcile regional, political, religious, and ethnic differences well enough not only to survive, but to prosper. Arend Lijphart, analyzing the Dutch reconciliation between social and ideological fragmentation and a working democracy, concluded that the Dutch experience is one example where the idea of "separate but equal" has worked. (Blakely 1993:8–11)

4. Blakely points out that

   the presence of newcomers in the population has been a characteristic feature of Dutch society for centuries. Immigration has occurred with a persistence and on a scale which leaves few present Dutch families without at least one ancestor from abroad. A spectacular early example was the arrival in the sixteenth century of approximately 150,000 Flemings and 75,000 Huguenots fleeing religious intolerance in France. The same century also witnessed a stream of Sephardic Jews from Spain and Portugal escaping the Inquisition. In addition, by the end of the seventeenth century 5,000 of Amsterdam's 7,500 Jewish population were Ashkenazi. (Blakely 1993:8–9)

5. The Frank family, made famous by Anne Frank's diary (Frank 1952), in which she describes her experiences of hiding in Amsterdam during the Nazi occupation of the Netherlands in the Second World War, offers a more recent example of Dutch tolerance. The Frank family, German by birth, went to the Netherlands specifically because Otto Frank, Anne's father, believed the family would be able to work and live in a society such as that of the Netherlands, which tolerated religious diversity (Gies and Gold 1995).

6. Goudsblom describes the structure as follows:

   At each level, from the nursery school to the university, parents may choose among three sorts of schools: the neutral public school, the confessional Protestant school, and the confessional Roman Catholic school. . . . The public schools are administered by the municipal town boards, the confessional schools by private boards. As the supplier of the material means, the state imposes binding conditions upon all schools regarding . . . admittance, curriculum, degrees, number of teachers, qualifications of teachers, and so forth . . . differences occur primarily in the religious teachings . . . but the dissimilarities extend into other subjects as well. . . . more important than the varied cultural content of the teaching is the social isolation caused by segmental education. Every child spends a large part of the day in an insular environment where contacts are virtually restricted to members of the same confessional or nonconfessional bloc (Goudsblom 1967:102–3).

7. In the Netherlands, students are compelled to attend school for ten years—eight years of elementary education and two subsequent years in one of the postelementary school programs. At age sixteen students must attend school two days a week and at seventeen one day a week is required. Elementary school begins at age four. Upon successfully completing eight grades, at about age twelve, students move on to one of five different types of educational programs, ranging from basic vocational education to university preparation.

Each program involves very different time commitments. The simplest vocational programs require two years of postelementary education, while the university preparation program requires six additional years of full-time study. The system became more flexible after a major reorganization in 1964, making the tracking less rigid and transfer between tracks, based on success or failure in a particular program, easier.

8. There is a similar discourse about the value of education—especially math and science education—in the Netherlands as well. At the same time, however, there is also wide recognition that not everyone wants or needs to excel in these areas to be a successful and valued adult.
9. Like many European countries, in the 1970s the Netherlands recruited so-called guest workers from poor nations to move to the Netherlands to work in jobs that could not be filled by Dutch workers. The Netherlands focused on rural Turkey and Morocco in recruiting such laborers.
10. Of course the meanings and usage of words develop and change over time. Both the dictionary of the Dutch language (de Vries et al. 1889) and a Dutch etymological dictionary (van Wijk 1912) note the relationship between *gewoon* and the verb *wonen*, meaning "to live" or "to reside," and give numerous examples of the possible uses of gewoon. The connection of gewoon to the verb wonen suggests the possibility of a further connotation. In Dutch, a past participle is made of regular verbs by adding "ge" to the front of the verb and changing the "en" ending to a "t" or a "d," making the past participle of the verb to live—wonen—*gewoond. Ik heb gewoond* means "I have lived." This construction suggests that gewoon may not just mean "ordinary" but may also have embedded in it the connotation of how things used to be. The dictionary of 1889 devotes more than four pages to the word.
11. The celebration of ordinariness in the Netherlands has a long history, one that is deeply tied to the dynamics of pillarization and embourgeoisement. In his social history of the Netherlands, Simon Schama analyzes a wide range of artwork from the Dutch masters of the Golden Age, the period during the seventeenth century when the Netherlands was a maritime trading power with colonies in Asia and the Caribbean. This period continues to have symbolic significance in Dutch social life today. Schama demonstrates that the presentation and content of paintings from this period celebrated ordinary, middle-class, everyday life (Schama 1988).

CHAPTER TWO: GENETICS AND GENETIC PRACTICE

1. Because much ethnographic and social studies of science work on genetics has focused on the United States and the United Kingdom, attention has only more recently begun to be paid to the ways local institutional structures shape genetic practices in different national contexts. This chapter offers such an analy-

sis for the Netherlands. Kaushik Sunder Rajan (2007) offers an analysis of the Indian context and Mike Fortun (2008) provides a study of Iceland.

2. My discussion in this section is drawn from knowledge developed during a genetics course intended for pre-medical students at the University of California, Berkeley, from fieldwork experiences, and from a genetics textbook (Mange and Mange 1990). My aim is to elaborate the prevailing understandings of genetics underlying the daily practice of genetics in the centers in which I conducted fieldwork as a means of explaining concepts and terms that arose repeatedly during my fieldwork.
3. The frequency of what are considered sex chromosome abnormalities has prompted biologist Ann Fausto-Sterling to argue that our contemporary understanding of biological sex—that there are two categories of sex, male and female, into which all individuals fit—is itself deeply social and ideological. See Fausto-Sterling (1993) for her analysis of the social, medical, and scientific work involved in fitting all people into the two available sex categories and her argument that there are more than two biological sexes.
4. This is a genetic disorder affecting numerous organ systems in affected individuals, including the skeleton, the lungs, the eyes and the heart and blood vessels, stemming from mutations in the molecule known as fibrillin (Sakai, Keene, and Engvall 1986). For an anthropological account of Marfan syndrome in the United States see Heath (1994, 1998).
5. This expectation was made particularly clear to me late one night when I accompanied a Dutch friend, Will, to the emergency room after he had been hit by a car on his bicycle. After our arrival we were asked to sit at a desk with a hospital staff person who requested information from my friend, which she then filled in on hospital forms. Although she was quite friendly and efficient she expressed clear disapproval after she asked Will for the name of his general practitioner and he told her that having recently returned from overseas travel, he did not have one.
6. At the time of my fieldwork non-citizens wishing to gain residence permits for the Netherlands had to be able to demonstrate that they had health insurance.
7. A degenerative muscle disorder, linked to chromosome 19, the symptoms for which vary greatly among those affected with cases ranging from very severe to extremely mild.
8. The research labs also use DNA from people outside of the country that comes to them via multiple routes. During my fieldwork in one lab a Ph.D. student was working with DNA from a pair of Brazilian siblings who lived four thousand kilometers from the clinic of the Brazilian physician through whom the DNA had come to the laboratory. The siblings' distance from the Brazilian clinic became a salient issue for the Dutch laboratory when they realized it would be

useful to have DNA from the parents of the siblings as well. When the question of whether it might be possible to procure such DNA arose, the Ph.D. student explained that she had contacted the Brazilian physician and was awaiting a response. She then described the problem of the distance of the people from whom the DNA was desired.

CHAPTER THREE: THE PRODUCTION OF ORDINARINESS

1. I refer to residents as junior clinicians and the other physicians as senior clinicians throughout. In the Netherlands a residency in medical genetics is four years.
2. The summary I have included here is my translation of a transparency used in a specific case presentation during my fieldwork at the clinic. I have left many of the original abbreviations in an attempt to maintain the flavor of the presentation, while in a few cases I have written out what was abbreviated in the original for the purposes of communicating to nonclinicians the intended meaning of the material presented on the transparency. For example, in the section on supplemental research where I have "chromosome research 46,XX" the original was "CO 46,XX." CO are the initials the clinicians use to signify *chromosome onderzoek*, which I translate as "chromosome research."
3. A pre-auricular tag is a mark, something like a dimple, on the face in front of the ear.
4. Stating that her height is in the p3 is shorthand for the third percentile, referring to where she falls on the pediatric growth charts.
5. As medical knowledge and technologies have made the inside of the body more visible, physicians at the academic hospital with which the genetics center is associated have developed a prenatal version of the audit. Every Wednesday at lunch an interdisciplinary group including a clinical geneticist, obstetricians and gynecologists, ultrasound technicians specializing in prenatal observation, and pediatric surgeons and neurologists meet to watch and discuss videotapes in order to analyze, diagnose, and follow up on prenatal ultrasounds in which irregularities have been observed.
6. These initials stand for *Oculo-Acoustic-Wervel*—the Dutch terminology for Cervico-Oculo-Acoustic, a spectrum of genetic disorders involving eye, ear, and spinal anomalies. The physician is speaking in English, but he uses the Dutch shorthand for the syndrome he is discussing. The code switching highlights the way physicians in the clinic constantly move between local and international contexts.
7. What might be considered personality traits are not infrequently considered as possible symptoms for various syndromes. "Cocktail party manners," for example, are considered a symptom in children with Williams syndrome.
8. Troy Duster has described race and genetics as a "conundrum without a solu-

tion." Debates about race, genetics, and biomedical practice are the subject of ongoing debate and controversy. For discussion of this debate see http://raceandgenomics.ssrc.org/.

9. Resistance to the taking of photos was rare and usually cause for comment. In fact, in the case of Ineke, which I elaborate below, her mother was reluctant at first to have the photos taken because she thought they were asking to take X-rays. When the physicians clarified that they wanted standard photos she immediately agreed. After the appointment the senior clinician involved told me that this was a problem in the Dutch language since in popular usage they use photo for both standard photographs and X-rays. In another case that a clinician presented she told us that the patient, an adult man, had agreed to have photos taken only on the condition that he got the negatives back once the genetic counseling was completed. She explained that he works in the medical library at the university, and she thought that his familiarity with the use of clinical photographs made him reluctant to have his own photos available for such use.
10. The purpose of photos of fetuses from terminated pregnancies or miscarriages was to diagnose the specific problem so as to be able to better predict the recurrence risk for the people involved.
11. These various technologies and techniques are described in chapter 2.
12. The issue of race and genomics is enormously complicated. Given the increasing knowledge of human biological variation, old as well as new biological definitions of race are entering (or reentering) medicine through genomics (see, for example, Braun 2002; Duster 2005; Goodman 2000; Kahn 2003, 2004; Lee 2003; Lee et al. 2001).
13. Keith Wailoo and Stephen Pemberton (2006) untangle the conflation of race, ethnicity, and the genetic conditions Tay-Sachs, cystic fibrosis, and sickle cell. Although, for example, many people in the United States, including physicians, view sickle cell as an "African" or "black" disease, it is, in fact, also found in Indian and Mediterranean populations, and it is virtually nonexistent in people from southern Africa. Rather than being an African or black disease, it is an artifact related to the presence of malaria. Sickle cell is a recessive condition, so one can be a carrier of one copy of the allele for the condition without being sick. Carrying two copies of the gene results in the painful and often debilitating condition known as sickle cell disease. Having one sickle cell allele turns out to make its bearer more resistant to malaria than those who do not carry one allele. Thus, carriers of one copy of this allele in regions with a high incidence of malaria throughout the world (West Africa, parts of India, the Mediterranean) were more likely to grow to adulthood and thus to reproduce than those who did not carry it. The fact that, especially in the United States, sickle cell is widely associated with African-Americans is a cultural artifact of slavers' hav-

ing kidnapped and enslaved West Africans. If they had kidnapped and enslaved Greeks, Americans would likely think of it as a Greek disease. Of course, such widespread assumptions can have significant clinical implications.

14. Rapp argues that "the notion that people with Down syndrome could be removed from the kinship nexus which was theirs 'by birth' and relocated inside their own separate tribe, caused me to wonder if we might describe their symbolic predicament as a 'kinship of affliction.' . . . The attribution of alien kinship does more than separate Down syndrome children from their genitors and genitrixes; it also provides an alternative kin group into which they can be placed" (Rapp 1994:81–82).

CHAPTER FOUR: BACKWARD AND BEAUTIFUL

1. The simplicity of this description hides the complicated process of in vitro fertilization, which typically requires the woman involved to undergo an extensive regime of hormone injections and surgical collection of enough ova that the laboratory will end up with enough fertilized embryos to work with. Generally speaking, gaining access to the sperm that will be involved in this process is a far simpler and less medicalized matter.
2. For an ethnographic treatment of preimplantation genetic diagnosis, see Sarah Franklin and Celia Roberts (2006).
3. Given national differences in the availability and cost of new biotechnologies, reproductive and medical tourism have emerged as global social phenomena. In this case Belgian social policy made preimplantation diagnosis available commercially, which it was not in the Netherlands.
4. I want to thank Stefan Helmreich for pointing this out to me.
5. The Amish play a similar role in the United States, although I would suggest that the contrast is more powerful in the Dutch case, intensified by the small size of both the country and its population and by a wider recognition of shared history.
6. There is great diversity of religious practices among the villages that make up the Bible Belt. At least one village is known for its practice of premarital sex, but it is not common to all of the villages popularly recognized as religious communities.
7. For example, Bourdieu (1977), Durkheim and Mauss (1963), Hallowell (1937), Lévi-Strauss (1963), Munn (1986), Rabinow (1989), Richards (1939), Tsing (1993).
8. Scholars who read the historical record as demonstrating Dutch Orthodox Reformed Calvinists' staunch opposition to capital accumulation argue that the Dutch case does not support Weber's thesis linking Calvinism and the rise of capitalism (Bainton 1936; Beekman 1935; Fanfani 1935; Hyma 1938; Robertson 1933).
9. The Dutch people I met frequently used the term *unlucky* to communicate that

a person had a genetic condition. For example, one senior citizen I spoke with told me she "had a sister who was unlucky" (*Ik had een zus die ongelukkig was*) because she was born with arm and leg anomalies. Physicians also often used this idiom, telling me that they wanted people to investigate their reproductive risks before they get an "unlucky child."

10. The potential value of such communities is highlighted by contemporary efforts to systematically mine them for genetic material, medical records, and family histories. Iceland offers a particularly vivid example of this phenomenon (Fortun 2001, 2008; Sigurdsson 2001; Rose 2001; Specter 1999).
11. The LOG is the monthly national meeting of medical geneticists described in chapter 2 and discussed in chapter 3.
12. There is perhaps an aesthetic pleasure in the symmetry of publishing under the name A. Zee.
13. For more information on cystic fibrosis, see http://www.cff.org.

CHAPTER FIVE: BOVINE ABOMINATIONS

1. It would seem, additionally, that it is no accident that Chinese scientists are attempting to clone a giant panda—an animal highly identified with the Chinese nation (*New York Times* 1999). In this regard the choice of a sheep for cloning in Scotland takes on added significance (Franklin 1997).
2. I do not intend to gloss Nazi science as an effort to create a super race. It was, as others have shown, both much more and much less than that. For analysis of Nazi science, see Kuhl (1994), Lifton and Markusen (1990), Muller-Hill (1988), and Proctor (1988, 1995). Many of the people I interviewed used the idea of Nazi science to talk about Nazi eugenics and the Nazi fantasy of producing a super race in talking about contemporary genetics. Hence, my use of these links follows popular Dutch conceptions of Nazi practices.
3. Interpretations of what happened in the Netherlands during the Second World War are hotly debated and intensely political. It would be impossible to provide a full analysis of the Second World War and the Netherlands here. For more general discussion of these issues in English, see Goudsblom (1967), Hirschfeld (1988 [1984]), Presser (1965), Shetter (1987), and van der Zee (1982).
4. The Dierenbescherming itself draws this comparison in its online history of the organization found on its Web site, www.dierenbescherming.nl.
5. In his ethnography of a small Dutch village (1947) that had a population of just over seven thousand people, Ivan Gadourek mentions the local chapter of the Dierenbescherming as one of several groups making up local civil society (1961:chap. 9).
6. These are precisely the terms scientists were using in 1997 to discuss the cloning of a sheep in Scotland. In one of her articles on the topic in the *New*

York Times, the science reporter Gina Kolata writes, "When a scientist whose goal is to turn animals into drug factories announced on Saturday in Britain that he and his team had cloned a sheep, the last practical barrier in reproductive technology was breached" (Kolata 1997:1). On the NBC *Today Show* (February 24, 1997), a newscaster reported that the cloned sheep was created as "part of an effort to find new ways of making drugs. They genetically alter livestock so the animals become machines that manufacture medicines to treat human diseases."

7. Also see Kruijt (1974) and Post (1989).
8. My thanks to Ann Anagnost for pointing this out to me.
9. In the case of euthanasia, this management comes from a policy of nonprosecution as long as physicians follow specific guidelines. In their book on Dutch legal culture, Blankenburg and Bruinsma report that these guidelines were announced by the "Procurators General after consulting the professional organization of medical practitioners" (1994:65).
10. At the time, however, Nazi ideas about eugenics reflected mainstream understandings of genetics that were operating widely among prominent scientists throughout the United States and Europe (Kevles 1995 [1985]; Paul 1995; Proctor 1988).
11. For a history of eugenics, see Kevles (1995 [1985]). For a history of Nazi science, see Proctor (1988).
12. I have translated the word *grens* here as "limit" because that translation makes the most sense in English. Grens is also the Dutch word for "border" or "frontier," including the kinds of borders that geographically define the nation.
13. The significance of the war in this regard is not limited to those people who themselves experienced it. Each time the topic came up with young people I interviewed, they told me they personally felt it was important to celebrate the holidays commemorating those who died during the war and the country's liberation. Rudie, an engineer in his early thirties, told me he could not imagine not going to his town square on May 4 to commemorate the war dead because he feels he owes those people something for his life. He then spoke about the experiences of his parents and his wife's father during the war. Tineke, a twenty-seven-year-old university student, sent me pictures of Canadian veterans marching through her town during the fiftieth anniversary of the country's liberation in 1995. She wrote, "I can't believe what these sweet old men did for us." The persisting significance of the war over time and across generations points to its place as an extremely important historical moment.
14. These aspects of Dutch social life are more fully elaborated in chapter 2. For more on the Dutch social structure known as verzuiling, see Goudsblom (1967), Kossmann-Putto and Kossman (1987), Kruijt (1974), Post (1989), and Shetter (1987).

15. This television documentary, *Beter Dan God* (Better than God) (Kayzer 1987), was at the time one of the most frequently watched television documentaries in Dutch history.
16. During my fieldwork I noticed that some nongovernmental organizations used concerns about immigration to encourage donations to NGOs working on development and relief efforts. They would suggest that providing resources to people in poor countries would encourage them to stay home rather than emigrate.
17. These policies, described in chapter 1, are rooted in the policies the Netherlands developed in the nineteenth century to accommodate the country's religious pluralism.
18. Also see Proctor (1991) for his analysis of "value-free" science.
19. The issue of whether nondirective counseling is ever possible is beyond the scope of my analysis here. For discussion of this question see Rapp (1999) and Rothman (1986, 1989, 1998).
20. Proctor's (1995) work on German physicians' participation in developing guidelines for "racial hygiene" makes this a problematic distinction. After all, Proctor argues, German physicians were intimately involved in constructing Jews, homosexuals, people with genetic abnormalities, and others as threats to public health and, therefore, medical problems.

EPILOGUE: ORDINARY GENOMES IN A GLOBALIZING WORLD

1. National biobanks are part of diverse and widespread efforts around the world to link DNA with family and medical histories as a means of contextualizing the knowledge about the structure of the human genome developed out of the Human Genome Project with the information that could help illuminate the relationship between structure and function in order to realize the promise of genetic interventions into human health. For further elaboration of this point, see Taussig (2005).
2. Sandra Soo-Jin Lee (2006) offers an excellent analysis, exploring in greater depth how the tropes surrounding biobanking translate diversity into "'racial identity' relinked to biological difference."
3. The genetic database policy was intended to facilitate widespread participation by employing a presumed consent method that allowed people to opt out. This meant that all citizens' DNA would be included in the database except that belonging to individuals who affirmatively chose not to participate. The ability to opt out was further constrained by the fact one's relatives' DNA tells something about one's own DNA, and the law did not allow people to prevent the inclusion of DNA from deceased relatives.

4. For example, one such poster was a picture of four laboratory test tubes, each containing an item (e.g., a fish) that was a symbol of the nation.
5. The foundation upon which stem cell research has developed comes from the knowledge and techniques developed with the cloning of Dolly the sheep (Wilmut et al. 1997). Dolly is significant because her creation involved what scientists had previously considered impossible: The resetting of a cell's developmental clock. Sarah Franklin (2001) points out that this event, in turn, called into question the possibility of the "biologically impossible" itself. Indeed, this biocultural achievement was transformative because it offered researchers new ways of thinking about nature and especially about the plasticity of life. Researchers quickly began envisioning a range of possible therapeutic interventions that the embryo technologies involved in Dolly could lead to, in particular the potential of embryos as sources of stem cells for therapeutic use and the possibility of what has come to be known as regenerative medicine.
6. For example, shortly after Bush's restrictions on human embryo research in 2001 *The Scientist* reported that a major University of California San Francisco (UCSF) stem cell researcher who had encountered difficulty in gaining federal support for his research was moving to Cambridge University. In the same article, "Scientists Seek Passports to Freer Environments," the magazine quoted a UCSF dean and vice chancellor as saying "if federal support for stem cell research is not forthcoming, the risk exists that talented scientists will leave academic centers to seek opportunities in the private sector, or even overseas. This would be a tragedy of the greatest proportion" (Brickley 2001:36).
7. Ironically, the primary dictionary definitions of *chimera* include "something totally unrealistic or impractical" (http://encarta.msn.com) and reference to the original use for "a fire-breathing she-monster in Greek mythology having a lion's head, a goat's body, and a serpent's tail" (http://www.m-w.com).
8. The Robert and Baylis (2003) article on chimeras and the crossing of species boundaries referenced here was a "Target Article" in the *American Journal of Bioethics*. As such, it was accompanied by twenty-four "Open Peer Commentaries" written by a range of historians, philosophers, ethicists, anthropologists, molecular biologists, and health policy analysts. Taken together these offer an interesting conversation about issues raised by the use of chimeras as research tools.

BIBLIOGRAPHY

Abraham-van der Mark, E. 1989. " 'I Don't Discriminate, but . . .': Dutch Stories about Ethnic Minorities." *Ethnologia Europeae* 19 (2): 169–84.

Agamben, Giorgio. 1998. *Homo Sacer: Sovereign Power and Bare Life*. Translated by D. Heller-Roazen. Stanford: Stanford University Press.

Alpers, Svetlana. 1983. *The Art of Describing: Dutch Art in the Seventeenth Century*. Chicago: University of Chicago Press.

Andrews, Lori. 1999. *The Clone Age*. New York: Henry Holt.

Angier, Natalie. 1995. " 'The Lady of the Flies' Dives into a New Pond." *New York Times*, 5 December, C1, C10.

Arensberg, Conrad. 1968. *The Irish Countryman*. Garden City: Natural History Press.

Arnason, E. 2000. "Genetic Homogeneity of Icelanders: Fact or Fiction?" *Nature Genetics* 25: 373–74.

Arnold, W. 2006. "Singapore Acts as Haven for Stem Cell Research." *New York Times*, 17 August, C1.

Associated Press. 2004. "Genetically Manipulated Bull Put to Sleep." Electronic document, http://my.netscape.com.

Bainton, H. 1936. "Changing Ideas and Ideals in the Sixteenth Century." *Journal of Modern History* 8 (4): 417–43.

Balibar, Étienne. 1991 [1988]. Preface to *Race, Nation, Class: Ambiguous Identities*, by Étienne Balibar and Immanuel Wallerstein, 1–13. London: Verso.

Barns, J. 1954. "Class and Committees in a Norwegian Island Parish." *Human Relations* 7:39–58.

Barth, Fredrik. 1969. *Ethnic Groups and Boundaries: The Social Organization of Culture Difference*. Boston: Little, Brown.

Beek, Leo. 1985. *Dutch Pioneers of Science*. Assen, Netherlands: Van Gorcum.

Beer, Gillian. 2000. *Darwin's Plots: Evolutionary Narrative in Darwin, George Eliot and Nineteenth-Century Fiction*. Cambridge: Cambridge University Press.

Beekman, E. H. M. 1935. *Katholicisme Calvinisme Kapitalisme*. Schiedam: Vox Romana.

Beidelman, T. O. 1980. "The Moral Imagination of the Kaguru: Some Thoughts on Tricksters, Translations, and Comparative Analysis." *American Ethnologist* 7 (1): 27–42.

Bell, Arthur. 1950. *Christian Huygens and the Development of Science in the Seventeenth Century*. London: Arnold.

Berger, John. 1973. *Ways of Seeing*. New York: Viking.

van Berkel, Klaas, Albert van Helden, and Lodewijk Palm, eds. 1999. *A History of Science in the Netherlands: Survey, Themes, and Reference*. Leiden: Brill.

Beurton, Peter J., Raphael Falk, and Hans Jorg Rheinberger. 2000. *The Concept of the Gene in Development and Evolution*. Cambridge: Cambridge University Press.

Bishop, J., and M. Waldholz. 1990. *Genome: The Story of the Most Astonishing Scientific Adventure of Our Time: The Attempt to Map All the Genes in the Human Body*. New York: Simon and Schuster.

Blakely, Allison. 1993. *Blacks in the Dutch World: The Evolution of Racial Imagery in a Modern Society*. Bloomington: Indiana University Press.

Blankenburg, Erhard, and Freek Bruinsma. 1994. *Dutch Legal Culture*. Deventer, Netherlands: Kluwer Law and Taxation.

Blok, A. 1975. *The Mafia of a Sicilian Village*. New York: Harper and Row.

Boas, Franz. 1928. *Anthropology and Modern Life*. New York: W. W. Norton.

Boer, P., F. Lagerwaard-Fijten, J. W. M. Marinissen, U. P. Steenhuis, and A. P. van der Zee. 1993. *Biologie Overal 6V*. Culemborg, Netherlands: Educatieve Partners Nederland.

Bourdieu, Pierre. 1967. "Systems of Education and Systems of Thought." *International Social Science Journal* 3 (19): 338–58.

——. 1977. *Outline of a Theory of Practice*. Cambridge: Cambridge University Press.

——. 1984. *Distinction: A Social Critique of the Judgment of Taste*. Translated by Richard Nice. Cambridge: Harvard University Press.

Bourdieu, Pierre, and J. Passeron. 1977. *Reproduction: In Education, Society and Culture*. Translated by Richard Nice. London: Sage.

van de Braak, H. 1993. *Homo Neerlandicus: Essay over "Wij Nederlanders."* Amersfoort: Enzo.

Braun, Lundy. 2002. "Race, Ethnicity, and Health: Can Genetics Explain Disparities?" *Perspectives in Biology and Medicine* 45 (2): 159–74.

Brendel, Carel. 2004. "So Much for Dutch Hospitality." *The Week* 4 (147) (12 March): 15.

Brickley, Peg. 2001. "Scientists Seek Passports to Freer Environments: Opposition to Federal Funding of Stem Cell Research Has Some Scientists Entertaining Plans to Move Abroad." *Scientist* 15 (16): 36.

Bulmer, Ralph. 1967. "Why Is the Cassowary Not a Bird? A Problem of Zoological Taxonomy among the Karam of the New Guinea Highlands." *Man* 2 (1): 5–25.

Canguilhem, Georges. 1991 [1966]. *The Normal and the Pathological.* New York: Zone.

Cantor, Charles. 1992. "The Challenges to Technology and Informatics." *The Code of Codes,* ed. D. Kevles and L. Hood, 98–111. Cambridge: Harvard University Press.

Caplan, Jane, and John C. Torpey. 2001. *Documenting Individual Identity: The Development of State Practices in the Modern World.* Princeton: Princeton University Press.

Carrier, James, ed. 1995. *Occidentalism: Images of the West.* New York: Oxford University Press.

Carson, Ronald A., and Mark A. Rothstein, eds. 2002. *Behavioral Genetics: The Clash of Culture and Biology.* Baltimore: Johns Hopkins University Press.

CDA [Christian Democratic Party]. 1992. *Genen en Grenzen: Een Christen-Democratische Bijdrage aan de Discussie Over de Gentechnologie: Rapport van een Commissie van het Wetenschappelijk Instituut Voor het CDA.* The Hague: het Wetenschappelijk Instituut voor het CDA.

Chapman, C. 1971. *Milocca: A Sicilian Village.* Cambridge, Mass.: Schenkman.

Claeson, Bjorn, Emily Martin, Wendy Richardson, Monica Schoch-Spana, and Karen-Sue Taussig. 1996. "Scientific Literacy? What It Is, Why It's Important, and Why Scientists Think We Don't Have It: The Case of Immunology and the Immune System." *Naked Science,* ed. Laura Nader, 101–16. New York: Routledge.

Cole, John. 1977. "Anthropology Comes Part-Way Home: Community Studies in Europe." *Annual Review of Anthropology* 6:349–78.

Cook-Deegan, Richard. 1994. *The Gene Wars: Science, Politics, and the Human Genome.* New York: W. W. Norton.

Cowan, Ruth. 1992. "Genetic Technology and Reproductive Choice: An Ethics for Autonomy." *The Code of Codes,* ed. D. Kevles and L. Hood, 244–63. Cambridge: Harvard University Press.

Crick, Francis. 1957. "On Protein Synthesis." *Symposium of the Society of Experimental Biology* 12:153.

——. 1981. *Life Itself.* New York: Simon and Schuster.

Crump, T. 1985. "Problems in the Local Kitchen: OR, Why the Pope Can't Speak Dutch." *Anthropology Today* 1 (6): 5–6.

Davis, Natalie Zemon. 1975. *Society and Culture in Early Modern France.* Stanford: Stanford University Press.

van Deursen, Arie. 1991. *Plain Lives in a Golden Age: Popular Culture, Religion and Society in Seventeenth-Century Holland.* Translated by Maarten Ultee. Cambridge: Cambridge University Press.

Dierenbescherming. N.d. *No To Genetic Manipulation of Animals.* The Hague: Dierenbescherming.

——. 1992. *Projectplan: Bio Technologie Bij Dieren*. The Hague: Dierenbescherming.
Dirks, Nicholas, ed. 1992. *Colonialism and Culture*. Ann Arbor: University of Michigan Press.
Douglas, Mary. 1966. *Purity and Danger*. London: Ark.
——. 1970. *Natural Symbols*. New York: Vintage.
Dreyfuss, R., and Dorothy Nelkin. 1992. "The Jurisprudence of Genetics." *Vanderbilt Law Review* 45 (2): 313–48.
Dreyfus, Hubert, and Paul Rabinow. 1983. *Michel Foucault: Beyond Structuralism and Hermeneutics*. Chicago: University of Chicago Press.
Duden, Barbara. 1991. *The Woman Beneath the Skin*. Cambridge: Harvard University Press.
Dumit, Joseph. 2003. *Picturing Personhood: Brain Scans and Biomedical Identity*. Princeton: Princeton University Press.
Durkheim, Émile, and Marcel Mauss. 1963. *Primitive Classification*. Chicago: University of Chicago Press.
Duster, Troy. 2005. "Race and Reification in Science." *Science* 307:1050–51.
Dwyer, K. 1982. *Moroccan Dialogues*. Baltimore: Johns Hopkins University Press.
Edwards, J., Sarah Franklin, E. Hirsch, F. Price, and Marilyn Strathern. 1993. *Technologies of Procreation: Kinship in the Age of Assisted Reproduction*. Manchester: Manchester University Press.
Ellemers, J. E. 1981. "The Netherlands in the Sixties and Seventies." *Netherlands Journal of Sociology* 17 (2): 113–35.
Epstein, Steven. 2007. *Inclusion: The Politics of Difference in Medical Research*. Chicago: University of Chicago Press.
Escobar, Arturo. 1994. "Welcome to Cyberia: Notes on the Anthropology of Cyberculture." *Current Anthropology* 35 (3): 211–32.
Evans-Pritchard, E. E. 1940. *The Nuer*. Oxford: Clarendon.
——. 1956. *Nuer Religion*. Oxford: Clarendon.
Fabian, J. 1983. *Time and the Other: How Anthropology Makes Its Object*. New York: Columbia University Press.
Fanfani, A. 1935. *Catholicism, Protestantism and Capitalism*. London: Sheed and Ward.
Fausto-Sterling, Anne. 1993. "The Five Sexes: Why Male and Female Are Not Enough." *The Sciences*, March–April, 20–24.
——. 1985. *Myths of Gender*. New York: Basic.
Feeley-Harnik, Gillian. 1999. "'Communities of Blood': The Natural History of Kinship in Nineteenth-Century America." *Comparative Studies in Society and History* 41 (2): 215–62.
Ford, Brian J. 1991. *The Leeuwenhoek Legacy*. Bristol, England: Biopress.
Fortun, Kim. 2001. *Advocacy after Bhopal: Environmentalism, Disaster, New Global Orders*. Chicago: University of Chicago Press.

Fortun, Michael. 2001. "Mediated Speculations in the Genomics Futures Market." *New Genetics and Society* 20 (2): 139–56.
——. 2008. *Promising Genomics: Iceland and deCODE Genetics in a World of Speculation.* Berkeley: University of California Press.
Foucault, Michel. 1979. *Discipline and Punish: The Birth of the Prison.* New York: Vintage.
——. 1980a. *The History of Sexuality.* Vol. 1: *An Introduction.* New York: Vintage.
——. 1980b. *Power/Knowledge: Selected Interviews and Other Writings.* Edited by C. Gordon. New York: Pantheon.
——. 1982. "Space, Knowledge and Power. Skyline (March)." Reprinted in *The Foucault Reader*, ed. Paul Rabinow, 239–56. New York: Random House.
Frank, Anne. 1952. *Diary of a Young Girl.* New York: Doubleday.
Franklin, Sarah. 1995. "Science as Culture, Cultures of Science." *Annual Review of Anthropology* 24:163–84.
——. 1997. *Embodied Progress: A Cultural Account of Assisted Conception.* London: Routledge.
——. 2000. "Life Itself." *Global Nature, Global Culture*, ed. Sarah Franklin, Celia Lury, and Jackie Stacey, 188–227. London: Sage.
——. 2001. "Culturing Biology: Cell Lines for the Second Millennium." *Health* 5 (3): 335–54.
——. 2007. *Dolly Mixtures: The Remaking of Genealogy.* Durham: Duke University Press.
Franklin, Sarah, and Margaret Lock, eds. 2003. *Remaking Life and Death: Towards an Anthropology of the Biosciences.* Santa Fe: School of American Research Press.
Franklin, Sarah, and Celia Roberts. 2006. *Born and Made: An Ethnography of Preimplantation Genetic Diagnosis.* Princeton: Princeton University Press.
Fukuyama, Francis. 2002. *Our Posthuman Future: Consequences of the Biotechnology Revolution.* New York: Farrar, Straus and Giroux.
Gadourek, I. 1961. *A Dutch Community: Social and Cultural Structure and Process in a Bulb-Growing Region in the Netherlands.* Groningen: J. B. Wolters.
Gannett, Lisa. 2002. "The Normal Genome in Twentieth-Century Evolutionary Thought: Studies in History and Philosophy of Science." *Part C* 34 (1):143–85.
Geertz, Clifford. 1973. *The Interpretation of Cultures.* New York: Basic.
——. 1983. *Local Knowledge.* New York: Basic.
Gies, Miep, and Allison Leslie Gold. 1995. *Anne Frank Remembered: The Story of the Woman Who Helped to Hide the Frank Family.* New York: Simon and Schuster.
Gilbert, Walter. 1992. "A Vision of the Grail." *The Code of Codes: Scientific and*

Social Issues in the Human Genome Project, ed. D. Kevles and L. Hood, 83–97. Cambridge: Harvard University Press.

Ginsburg, Faye, and Rayna Rapp. 2001. "Enabling Disability." *Public Culture* 13 (3): 533–56.

Goldstein, Judith. 1995. "Realism without a Human Face." *Spectacles of Realism: Body, Gender, Genre*, ed. Margaret Cohen and Christopher Prendergast, 66–89. Minneapolis: University of Minnesota Press.

Good, Mary-Jo. 1995. "Cultural Studies of Biomedicine: An Agenda for Research." *Social Science and Medicine* 41 (4): 461–73.

Goodman, Alan. 2000. "Why Genes Don't Count (for Racial Differences in Health)." *American Journal of Public Health* 90 (11): 1699–1702.

Goodman, Alan, Deborah Heath, and M. Susan Lindee, eds. 2003. *Genetic Nature/Culture: Anthropology and Science Beyond the Two-Culture Divide.* Berkeley: University of California Press.

Goudsblom, J. 1967. *Dutch Society*. New York: Random House.

Gupta, Akhil, and James Ferguson. 1997. *Anthropological Locations: Boundaries and Grounds of a Field Science.* Berkeley: University of California Press.

Haas, Scott. 2004. *Are We There Yet?* New York: Plume/Penguin.

Habermas, Jürgen. 2001. "On the Way to Liberal Eugenics? The Dispute over the Ethical Self-Understanding of the Species." Paper presented at the Colloquium on Law, Philosophy, and Political Theory, New York University School of Law, 25 October and 1 November. New York.

Hacking, Ian. 1990. *The Taming of Chance.* Cambridge: Cambridge University Press.

Hall, Stuart. 1990. "James Watson and the Search for Biology's 'Holy Grail.'" *Smithsonian* 20 (February): 41–49.

Hallowell, Irving. 1937. "Temporal Orientation in Western Civilization." *American Anthropologist* 39:647–70.

Haraway, Donna. 1991. "Situated Knowledges: The Science Question in Feminism and the Privilege of Partial Perspective." *Simians, Cyborgs, and Women: The Reinvention of Nature.* New York: Routledge.

——. 1997. *Modest Witness@Second Millennium.FemaleMan©MeetsOnco Mouse™: Feminism and Technoscience.* New York: Routledge.

——. 2003. "For the Love of a Good Dog: Webs of Action in the World of Dog Genetics." *Race, Nature, and the Politics of Difference*, ed. Donald Moore, Jake Kosek, and Anand Pandian, 254–95. Durham: Duke University Press.

Harvey, David. 1989. *The Condition of Postmodernity: An Enquiry into the Origins of Cultural Change.* Oxford: Blackwell.

Heath, Deborah. 1994. "Articulatory Biotechnologies." Paper presented at the Society for Social Study of Science Meetings, New Orleans, 13 October.

——. 1998. "Locating Genetic Knowledge: Picturing Marfan Syndrome and

Its Traveling Constituencies." *Science, Technology and Human Values* 23 (1): 71–97.

Heath, Deborah, and Paul Rabinow, eds. 1993. "Bio-Politics: The Anthropology of the New Genetics and Immunology." *Journal of Culture, Medicine, and Psychiatry* 17 [special issue].

Heath, Deborah, Rayna Rapp, and Karen-Sue Taussig. 2004. "Genetic Citizenship." *A Companion to the Anthropology of Politics*, ed. D. Nugent and J. Vincent, 152–67. London: Blackwell.

Helmreich, Sefan. 1998. *Silicon Second Nature: Culturing Artificial Life in a Digital World.* Berkeley: University of California Press.

Hirschfeld, G. 1988 [1984]. *Nazi Rule and Dutch Collaboration.* Translated by L. Willmot. New York: Berg.

Hood, Leroy. 1992. "Biology and Medicine in the Twenty-First Century." *Code of Codes: Scientific and Social Issues in the Human Genome Project*, ed. D. Kevles and L. Hood, 136–63. Cambridge: Harvard University Press.

Hubbard, Ruth, and Elijah Wald. 1993. *Exploding the Gene Myth: How Genetic Information is Produced and Manipulated by Scientists, Physicians, Employers, Insurance Companies, Educators, and Law Enforcers.* Boston: Beacon.

Huerta, Robert D. 2003. *Giants of Delft: Johannes Vermeer and the Natural Philosophers: The Parallel Search for Knowledge during the Age of Discovery.* Lewisburg, Penn.: Bucknell University Press.

Hyma, A. 1938. "Calvinism and Capitalism in the Netherlands, 1555–1700." *Journal of Modern History* 10 (3): 321–43.

Inda, Jonathan X., and Renato Rosaldo, eds. 2001. *Anthropology of Globalization.* Malden, Mass.: Blackwell.

Jackson, Peter. 1992. "Construction of Culture, Representations of Race: Edward Curtis's 'Way of Seeing.'" *Inventing Places: Studies in Cultural Geography*, ed. K. Anderson and F. Gale, 89–106. Melbourne: Longman Cheshire.

Jasanoff, Sheila. 2005. *Designs on Nature: Science and Democracy in Europe and the United States.* Princeton: Princeton University Press.

de Jong, Louis. 1969–88. *Het Koninkrijk der Nederlanden in de Tweede Wereldoorlog.* (Rijksinstituut voor Oorlogsdocumentatie). The Hague: Staatsuitgeverij.

de Jong, Perro. 1994. "Dutch Fear Loss of Tolerance." Radio Netherlands via BBC News, 3 November. http://news.bbc.co.uk

Jordanova, Ludmilla. 1989. *Sexual Visions: Images of Gender in Science and Medicine Between the Eighteenth and Twentieth Centuries.* Madison: University of Wisconsin Press.

Judson, H. F. 1979. *The Eighth Day of Creation: Makers of the Revolution in Biology.* New York: Simon and Schuster.

Kahn, Jonathan. 2003. "Getting the Numbers Right: Statistical Mischief and Racial

Profiling in Heart Failure Research." *Perspectives in Biology and Medicine* 46 (4): 473–83.
——. 2004. "How a Drug Becomes Ethnic: Law, Commerce, and the Production of Racial Categories in Medicine." *Yale Journal of Health Policy, Law, and Ethics* 4:1–46.
Kaldenbach, H. 1994. *Doe Maar Gewoon: 99 Tips Voor Het Omgaan met Nederlanders*. Amsterdam: Prometheus.
Kay, Lilly. 2000. *Who Wrote the Book of Life?* Stanford: Stanford University Press.
Kayzer, Wim. 1987. *Beter Dan God*. Hilversum: VPRO.
Keller, Evelyn Fox. 1983. *A Feeling for the Organism: The Life and Work of Barbara McClintock*. New York: W. H. Freeman.
——. 1992. "Nature, Nurture, and the Human Genome Project." *Code of Codes: Scientific and Social Issues in the Human Genome Project*, ed. Daniel Kevles and Leroy Hood, 281–99. Cambridge: Harvard University Press.
——. 1995. *Refiguring Life: Metaphors of Twentieth-Century Biology*. New York: Columbia University Press.
——. 2000. *The Century of the Gene*. Cambridge: Harvard University Press.
——. 2002. *Making Sense of Life*. Cambridge: Harvard University Press.
Kevles, Daniel. 1995 [1985]. *In the Name of Eugenics*. Cambridge: Harvard University Press.
Kevles, Daniel, and Leroy Hood, eds. 1992. *The Code of Codes: Scientific and Social Issues in the Human Genome Project*. Cambridge: Harvard University Press.
Kirschner, Suzanne R. 1999. "From Flexible Bodies to Fluid Minds: An Interview with Emily Martin." *Ethos* 27 (3): 247–82.
Klep, C. 1994. "The Second World War and the Netherlands." Lecture presented for the Summer Programme in Dutch Society and Culture, Faculty of Letters, University of Utrecht, 29 July, Utrecht, Netherlands.
Köhler, W. 1991. "Kalfje Geboren uit Cel Met Genetische Codering van Mens." *NRC Handelsblad*, 2 May.
——. 1994. "Nutricia Stond Heimelijk aan de Wieg van Stier Herman." *NRC Handelsblad* 2 June, 1.
Kolata, Gina. 1993. "Scientist Clones Human Embryos, and Creates an Ethical Challenge." *New York Times*, 24 October, A1.
——. 1997 "With Cloning of a Sheep, the Ethical Ground Shifts." *New York Times*, 24 February, A1.
Kondo, Ikuko, Shigetoshi Nagataki, and Nobuo Miyagi. 1990. "The Cohen Syndrome: Does Mottled Retina Separate a Finnish and a Jewish Type?" *American Journal of Medical Genetics* 37 (1): 109–13.
Kossman-Putto, J., and E. Kossman. 1987. *The Low Countries*. Rekkem, Belgium: Flemish-Netherlands Foundation.

Kruijt, J. P. 1974. "The Netherlands: The Influence of Denominationalism on Social Life and Organizational Patterns." *Consociational Democracy: Political Accommodation in Segmented Societies*, ed. K. D. McRae, 128–36. Toronto: McClelland and Stewart.

Kuhl, Stefan. 1994. *The Nazi Connection*. New York: Oxford University Press.

Kuhn, Thomas S. 1962. *The Structure of Scientific Revolutions*. Chicago: University of Chicago Press.

Kuriyama, Shigehisa. 1999. *The Expressiveness of the Body: The Divergence of Greek and Chinese Medicine*. New York: Zone.

Langford, Jean M. 2002. *Fluent Bodies: Ayurvedic Remedies for Postcolonial Imbalance (Body, Commodity, Text)*. Durham: Duke University Press.

Latour, Bruno. 1987. *Science In Action: How to Follow Scientists and Engineers Through Society*. Cambridge: Harvard University Press.

——. 1988. *The Pasteurization of France*. Cambridge: Harvard University Press.

——. 1990. "Postmodern? No, Simply Amodern! Steps towards an Anthropology of Science." *Studies in the History and Philosophy of Science* 21 (1): 145–71.

——. 1993. *We Have Never Been Modern*. Translated by Catherine Porter. Cambridge: Harvard University Press.

Lechner, Frank, and John Boli. 2004. *The Globalization Reader*. Malden, Mass.: Blackwell.

Lee, Sandra Soo-Jin. 2003. "Race, Distributive Justice and the Promise of Pharmacogenomics: Ethical Considerations." *American Journal of Pharmacogenomics* 3 (6): 385–92.

——. 2006. "Biobanks of a 'Racial Kind': Mining for Difference in the New Genetics." *Patterns of Prejudice* 40 (4–5): 443–60.

Lee, Sandra Soo-Jin, Joanna Mountain, and Barbara Koenig. 2001. "The Meanings of 'Race' in the New Genomics: Implications for Health Disparities Research." *Yale Journal of Health Policy, Law, and Ethics* 1:33–75.

Lévi-Strauss, Claude. 1963. *Structural Anthropology*. New York: Random House.

Lewontin, Richard. 1991. *Biology as Ideology: The Doctrine of DNA*. New York: Harper Perennial.

——. 1999. "Forget Privacy: Iceland Sells Citizens' Personal, Health, Genetic Records." *Cleveland Plain Dealer*, 27 January, 11B.

——. 2000. *The Triple Helix: Gene, Organism, and Environment*. Cambridge: Harvard University Press.

——. 2001. *It Ain't Necessarily So: The Dream of the Human Genome and Other Illusions*. New York: New York Review of Books.

Lifton, Robert J., and Eric Markusen. 1990. *The Genocidal Mentality*. New York: Basic.

Lindee, M. Susan. 2005. *Moments of Truth in Genetic Medicine*. Baltimore: Johns Hopkins University Press.

Lindee, M. Susan, Alan Goodman, and Deborah Heath. 2003. "Introduction: Anthropology in the Age of Genetics: Practice, Discourse, and Critique." *Genetic Nature/Culture: Anthropology and Science Beyond the Two-Culture Divide*, ed. Alan Goodman, Deborah Heath, and M. Susan Lindee, 1–19. Berkeley: University of California Press.

Lock, Margaret. 1993. *Encounters with Aging: Mythologies of Menopause in Japan and North America*. Berkeley: University of California Press.

——. 2001. *Twice Dead: Organ Transplantation and the Reinvention of Death*. Berkeley: University of California Press.

——. 2004. *On the Unfashionability of Certain "Genetic" Diseases*. Toronto: Canadian Institutes of Health Research Conference.

Lorwin, V. R. 1974. "Segmented Pluralism: Ideological Cleavages and Political Cohesion in the Smaller European Democracies." *Consociational Democracy: Political Accommodation in Segmented Societies*, ed. K. D. McRae. Toronto: McClelland and Stewart.

Mange, A., and E. Mange. 1990. *Genetics: Human Aspects*. Los Angeles: Sinauer.

Marcus, George. 1995. "Ethnography In/Of the World System: The Emergence of Multi-Sited Ethnography." *Annual Review of Anthropology* 24:95–117.

Marks, Jonathan. 2002. *What It Means to be 98% Chimpanzee: Apes, People, and Their Genes*. Berkeley: University of California Press.

Martin, Emily. 1987. *The Woman in the Body: A Cultural Analysis of Reproduction*. Boston: Beacon.

——. 1990a. "The Ideology of Reproduction: The Reproduction of Ideology." *Uncertain Terms: Negotiating Gender in American Culture*, ed. F. Ginsburg and A. L. Tsing, 300–314. Boston: Beacon.

——. 1990b. "Toward an Anthropology of Immunology: The Body as Nation State." *Medical Anthropology Quarterly* 4 (4): 410–26.

——. 1991. "The Egg and The Sperm: How Science Has Constructed a Romance Based on Stereotypical Male-Female Roles." *Signs: Journal of Women in Culture and Society* 16 (3): 485–501.

——. 1994. *Flexible Bodies: Tracking Immunity in American Culture from the Days of Polio to the Age of AIDS*. Boston: Beacon.

——. 1997. "Anthropology and the Cultural Study of Science: From Citadels to String Figures." *Anthropological Locations: Boundaries and Grounds of a Field Science*, ed. Akhil Gupta and James Ferguson, 131–46. Berkeley: University of California Press.

——. 2007. *Bipolar Expeditions: Mania and Depression in American Culture*. Princeton: Princeton University Press.

Martin, Emily, Laura Oaks, A. van der Straten, and Karen-Sue Taussig. 1997. "AIDS, Knowledge, and Discrimination in the Inner City: An Anthropological Analysis of Experiences of Injection Drug Users." *Cyborgs and Citadels: Anthro-*

pological Interventions into Techno-Humanism, ed. Gary Downy, Joseph Dumit, Sharon Traweek, 49–66. Santa Fe: School of American Research Press.

Mauss, Marcel. 1979. "The Notion of Body Techniques." *Sociology and Psychology: Essays*, 95–119. London: Routledge.

McClintock, Anne. 1995. *Imperial Leather: Race, Gender, and Sexuality in the Colonial Conquest.* New York: Routledge.

Mendel, Gregor. 1865/1866. "Versuche Über Pflanzenhybriden." *Verhandlungen des Naturforschenden Vereines in Brunn* 4:3–47. [The original paper was published in 1865, but the issue appeared in 1866. A translation by W. Bateson is printed in *Evolutionary Genetics. Benchmark Papers in Genetics*, vol. 8, ed. D. Jameson. (Stroudsburg, Penn.: Dowden, Hutchinson and Ross).]

Middendorp, C. P. 1991. *Ideology in Dutch Politics: The Democratic System Reconsidered 1970–1985*. Assen: Van Gorcum.

Morgan, Lynn. 1989. "When Does Life Begin? A Cross-Cultural Perspective on the Personhood of Fetuses and Young Children." *Abortion Rights and Fetal Personhood*, ed. E. Doerr and J. Prescott, 97–114. Long Beach: Centerline.

Morgan, Lewis Henry. 1870. *Systems of Consanguinity and Affinity of the Human Family. Smithsonian Contributions to Knowledge*, no. 17. Washington: Smithsonian Institution Press.

Muller-Hill, Benno. 1988. *Murderous Science: Elimination by Scientific Selection of Jews, Gypsies, and Others, Germany 1933–1945*. Translated by George Fraser. Oxford: Oxford University Press.

Munn, Nancy. 1986. *Fame of Gawa*. Cambridge: Cambridge University Press.

NBC. 1997. "Today." 24 February.

Nelkin, Dorothy, and Susan Lindee. 1996. *The DNA Mystique: The Gene as Cultural Icon*. New York: W. H. Freeman.

New York Times. 1999. "Chinese Scientists Try to Clone Giant Panda," 25 June, D5.

Noordman, Jan. 1989. *Om de Kwaliteit van het Nageslacht: Eugenetica in Nederland 1900–1950*. Nijmegen: SUN.

——. 1994. "Eugenics and the Mental Health Movement in the Netherlands 1930–1950." *Population and Family in the Low Countries*, ed. H. van den Brekel and F. Deven, 107–23. Amsterdam: Kluwer.

Novas, C., and Nikolas Rose. 2000. "Genetic Risk and the Birth of the Somatic Individual." *Economy and Society* 29 (4): 485–513.

Ong, Aihwa. 1995. "Making the Biopolitical Subject: Cambodian Immigrants, Refugee Medicine and Cultural Citizenship in California." *Social Science and Medicine*, 1–15.

——. 1996. "Cultural Citizenship as Subject Making." *Current Anthropology* 37 (5): 737–62.

——. 1999. *Flexible Citizenship: The Cultural Logics of Transnationality*. Durham: Duke University Press.

——. 2005. "Ecologies of Expertise: Assembling Flows, Managing Citizenship." *Global Assemblage Technology, Politics, and Ethics as Anthropological Problems*, ed. A. Ong and S. Collier, 337–53. London: Blackwell.
——. 2006. "A Canary in the Asian Mine: Ethicalization at Multiple Scales." Paper presented at the conference Ethical Worlds of Stem Cell Research, University of California, San Francisco, 29 September.
Parens, Erik, Audrey R. Chapman, and Nancy Press, eds. 2005. *Wrestling with Behavioral Genetics: Science, Ethics, and Public Conversation*. Baltimore: Johns Hopkins University Press.
de Pater, J., and C. Bos. 1993. "Prenatale Diagnostiek." *Clinical Genetics: Impact on Individual and Society; Present Status and Future*. Proceedings from the Symposium ter gelegenheid van het derde lustrum van de Stichting Klinisch Genetisch Centrum Utrecht, Utrecht, Netherlands, September.
Paul, Diane. 1995. *Controlling Human Heredity, 1865 to the Present*. Atlantic Highlands, N.J.: Humanities Press.
Petryna, Adriana. 2002. *Life Exposed*. Princeton: Princeton University Press.
Poovey, Mary. 1988. *Uneven Developments: The Ideological Work of Gender in Mid-Victorian England*. Chicago: University of Chicago Press.
Post, Harry. 1989. *Pillarization: An Analysis of Dutch and Belgian Society*. Brookfield, Vt.: Gower.
Presser, J. 1965. *Ashes in the Wind: The Destruction of Dutch Jewry*. Translated by A. Pomerans. London: Souvenir.
Proctor, Robert. 1988. *Racial Hygiene: Medicine Under the Nazis*. Cambridge: Harvard University Press.
——. 1991. *Value-Free Science? Purity and Power in Modern Knowledge*. Cambridge: Harvard University Press.
——. 1995. "The Destruction of 'Lives Not Worth Living.'" *Deviant Bodies*, ed. Jennifer Terry and Jacqueline Urla, 170–96. Bloomington: Indiana University Press.
Rabinow, Paul. 1989. *French Modern*. Cambridge: MIT Press.
——. 1992a. "Artificiality and Enlightenment: From Sociobiology to Biosociality." *Incorporations: Zone 6*, ed. J. Crary and S. Kwinter. New York: Urzone.
——. 1992b. "Severing the Ties: Fragmentation and Redemption in Late Modernity." *The Anthropology of Science and Technology: Knowledge and Society 9 D.*, ed. David Hess and Linda Layne. Greenwich, Conn.: JAI.
——. 1999. *French DNA: Trouble in Purgatory*. Chicago: University of Chicago Press.
Rajan, Kaushik Sunder. 2007. *Biocapital: The Constitution of Postgenomic Life*. Durham: Duke University Press.
Rapp, Rayna. 1991. "Moral Pioneers: Women, Men and Fetuses on a Frontier of Reproductive Technology." *Gender at the Crossroads of Knowledge*, ed. Micaela di Leonardo, 383–95. Berkeley: University of California Press.

———. 1994a. "Risky Business: Genetic Counseling in a Shifting World." *Articulating Hidden Histories*, ed. Jane Schneider and Rayna Rapp, 175–89. Berkeley: University of California Press.

———. 1994b. "Women's Responses to Prenatal Diagnosis: A Sociocultural Perspective on Diversity in Women and Prenatal Testing." *Facing the Challenges of Genetic Technology*, ed. Karen Rothenberg and Elizabeth Thomson, 219–33. Columbus: Ohio State University Press.

———. 1999. *Testing Women, Testing the Fetus*. New York: Routledge.

Reardon, Jenny. 2004. *Race to the Finish: Identity and Governance in an Age of Genomics*. Princeton: Princeton University Press.

Richards, Audrey I. 1939. *Land, Labour and Diet in Northern Rhodesia: An Economic Study of the Bemba Tribe*. Oxford: Oxford University Press.

Ridley, Matt. 1999. *Genome: The Autobiography of a Species in 23 Chapters*. New York: Harper Collins.

Rivers, W. H. R. 1922. "The Psychological Factor." *Essays on the Depopulation of Melanesia*, ed. W. H. R. Rivers, 84–113. Cambridge: Cambridge University Press.

Robert, Jason Scott, and Françoise Baylis. 2003. "Crossing Species Boundaries." *American Journal of Bioethics* 3 (3): 1–13.

Roberts, Lissa. 1999. "Going Dutch: Situating Science in the Dutch Enlightenment." *Sciences in Enlightened Europe*, ed. William Clark et al., 350–88. Chicago: University of Chicago Press.

Robertson, H. M. 1933. *Aspects of the Rise of Economic Individualism: A Criticism of Max Weber and His School*. Cambridge: Cambridge University Press.

Rose, Hilary. 2001. "Gendered Genetics in Iceland." *New Genetics and Society* 20 (1): 119–38.

Rose, Nikolas. 2006. *The Politics of Life Itself: Biomedicine, Power, and Subjectivity in the Twenty-First Century*. Princeton: Princeton University Press.

Rose, Nikolas, and Carlos Novas. 2005. "Biological Citizenship." *Global Assemblages: Technology, Politics, and Ethics as Anthropological Problems*, ed. Aihwa Ong and Stephen J. Collier, 439–63. Oxford: Blackwell.

Rothman, Barbara Katz. 1986. *The Tentative Pregnancy: Prenatal Diagnosis and the Future of Motherhood*. New York: W. W. Norton.

———. 1989. *Recreating Motherhood: Ideology and Technology in a Patriarchal Society*. New York: W. W. Norton.

———. 1999. *Genetic Maps and Human Imaginations*. New York: W. W. Norton.

Roxburgh, Angus. 2004. "Dutch Are 'Polarized,' Says Report." Electronic document, http://newsvote.bbc.co.uk.

Sahlins, Marshall. 1976. *The Use and Abuse of Biology: An Anthropological Critique of Sociobiology*. Ann Arbor: University of Michigan Press.

Sakai, L. Y., D. R. Keene, and E. Engvall. 1986. "Fibrillin, a New 350-kD Glycopro-

tein, Is a Component of Extracellular Microfibrils." *Journal of Cell Biology* 103:2499–2509.
Schaffer, Simon. 1989. "The Glorious Revolution and Medicine in Britain and the Netherlands." *Notes and Records of the Royal Society of London* 43 (2): 167–90 [special issue: Science and Civilization under William and Mary].
Schama, Simon. 1988. *The Embarrassment of Riches: An Interpretation of Dutch Culture in the Golden Age*. Berkeley: University of California Press.
Schellekens, Huub, and R. Visser. 1987. *De Genetische Manipulatie*. Amsterdam: Meulenhoff.
Scheper-Hughes, Nancy, and Margaret Lock. 1987. "The Mindful Body: A Prolegomenon to Future Work in Medical Anthropology." *Medical Anthropology Quarterly* 1 (1): 6–41.
Schiebinger, Londa. 1993. *Nature's Body*. Boston: Beacon.
Schneider, Jane, and Peter Schneider. 1976. *Culture and Political Economy in Western Sicily*. New York: Academic.
Shafir, Gershon, ed. 1998. *The Citizenship Debates: A Reader*. Minneapolis: University of Minnesota Press.
Shetter, William Z. 1987. *The Netherlands in Perspective: The Organization of Society and Environment*. Leiden: Martinus Nijhoff.
Shields, Rob. 1991. *Places on the Margin: Alternative Geographies of Modernity*. London: Routledge Chapman Hall.
Sigurdsson, Skuli. 2001. "Yin-Yang Genetics, or the HSD DECODE Controversy." *New Genetics and Society* 20 (2): 103–17.
Simons, Marlise. 1998. "Bosnia Massacre Mars Do-Right Self-Image the Dutch Hold Dear." *New York Times*, 13 September, 4.
Simonutti, Luisa. 1999. "Religion, Philosophy, and Science: John Locke and Limborch's Circle in Amsterdam." *Everything Connects: In Conference with Richard H. Popkin: Essays in his Honor*, ed. James E. Force and David S. Katz, 295–324. Leiden: Brill.
Specter, Michael. 1999. "Iceland Decoded." *New Yorker*, 18 January, 40–51.
Stephenson, Peter. 1989. "Going to McDonald's in Leiden: Reflections on the Concept of Self and Society in the Netherlands." *Ethos* 17 (2): 226–47.
——. 1990. "Does New Land Mean New Lives? Symbolic Contradiction and the Unfinished Reclamation of the Markerwaard." *Etnofoor* 3 (2): 17–31.
Stewart, Susan. 1993. *On Longing: Narratives of the Miniature, the Gigantic, the Souvenir, the Collection*. Durham: Duke University Press.
Stolberg, Sheryl. 1999. "The Biotech Death of Jesse Gelsinger." *New York Times*, Sunday Magazine, 28 November.
Stoler, Ann Laura. 2002. *Carnal Knowledge and Imperial Power: Race and the Intimate in Colonial Rule*. Berkeley: University of California Press.
Strathern, Marilyn. 1992a. *After Nature: English Kinship in the Late Twentieth Century*. Cambridge: Cambridge University Press.

——. 1992b. *Reproducing the Future: Anthropology, Kinship, and the New Reproductive Technologies.* New York: Routledge.

——. 1993. "Introduction: A Question of Context." *Technologies of Procreation: Kinship in the Age of Assisted Conception*, ed. Jeanette Edwards, Sarah Franklin, Eric Hirsch, Frances Price, and Marilyn Strathern, 1–19. Manchester: Manchester University Press.

Suzuki, D., and P. Knudtson. 1990 [1989]. *Genethics: The Clash between the New Genetics and Human Values.* Cambridge: Harvard University Press.

Taussig, Karen-Sue. 2005. "The Molecular Revolution in Medicine: Promise, Reality, and Social Organization." *Complexities: Beyond Nature and Nurture*, ed. S. McKinnon and S. Silverman, 223–47. Chicago: University of Chicago Press.

Taussig, Karen-Sue, Rayna Rapp, and Deborah Heath. 2003. "Flexible Eugenics: Technologies of the Self in the Age of Genetics." *Genetic Nature/Culture: Anthropology in the Age of Genetics, Genetics in the Age of Anthropology*, ed. A. Goodman, S. Lindee, and D. Heath, 58–76. Berkeley: University of California Press.

Taylor, Janelle. 1992. "The Public Fetus and the Family Car: From Abortion Politics to a Volvo Advertisement." *Public Culture* 4 (2): 167–83.

——. 2003. "The Story Catches You and You Fall Down." *Medical Anthropology Quarterly* 17 (2): 159–81.

Torpey, John C. 2000. *The Invention of the Passport: Surveillance, Citizenship, and the State.* Cambridge: Cambridge University Press.

Torres, Rodolfo, Jonathan X. Inda, and Louis F. Mirón, eds. 1999. *Race, Identity, and Citizenship: A Reader.* Malden, Mass.: Blackwell.

Traweek, Sharon. 1988. *Beamtimes and Lifetimes: The World of High Energy Physicists.* Cambridge: Harvard University Press.

Tringe, Michael. 2001. "Icelandic DNA: DeCODE Genetics and the Social Practice of Biotechnology." Senior Thesis, Department of the History of Science, Harvard University.

Tsing, Anna. 1993. *In the Realm of the Diamond Queen.* Princeton: Princeton University Press.

——. 2000. "The Global Situation." *Cultural Anthropology* 15 (3): 327–60.

——. 2005. *Friction: An Ethnography of Global Connection.* Princeton: Princeton University Press.

Tutton, Richard. 2007. "Constructing Participation in Genetic Databases: Citizenship, Governance, and Ambivalence." *Science, Technology and Human Values* 32 (2): 172–95.

Turner, Victor. 1995 [1969]. *The Ritual Process: Structure and Anti-Structure.* Hawthorne, N.Y.: Aldine de Gruyter.

van der Veen, D. J. 1994. "The Making of a Modern State: Pillarization, Social Organisation and the Welfare State." Lecture presented for the Summer Programme in Dutch Society and Culture, Faculty of Letters, University of Utrecht, 13 July, Utrecht, Netherlands.

Vereniging Samenwerkende Ouder- en Patiëntenorganisaties (VSOP). 1987. *Different and Yet the Same*. Baarn, Netherlands: VSOP.

de Vries, M., E. Verwijs, P. J. Cosijn, A. Kluyver, A. Beets, and J. W. Muller. 1889. *Woordenboek der Nederlandsche Taal*. The Hague: M. Nijhoff, A. W. Sijthoff.

Wailoo, Keith, and Stephen Pemberton. 2006. *The Troubled Dream of Genetic Medicine: Ethnicity and Innovation in Tay-Sachs, Cystic Fibrosis, and Sickle Cell Disease*. Baltimore: Johns Hopkins University Press.

Wallerstein, Immanuel. 1974. *The Modern World System: Capitalist Agriculture and the Origins of the European World Economy in the 16th Century*. New York: Academic.

———. 1980. *The Modern World System II: Mercantilism and the Consolidation of the European World Economy, 1600–1750*. New York: Academic.

Watson, James. 1968. *The Double Helix*. New York: Atheneum.

———. 1990. "The Human Genome Project: Past, Present, and Future." *Science* 248 (6 April): 44–49.

Watson, J., and F. Crick. 1953. "Molecular Structure of Nucleic Acid: A Structure for Deoxyribonucleic Acid." *Nature* 171:737–38.

Weber, Max. 1958 [1904]. *The Protestant Ethic and the Spirit of Capitalism*. New York: Charles Scribner's Sons.

Weiner, Annette. 1993. "Culture and Our Discontents." Presidential Address. 92nd Annual Meetings of the American Anthropological Association, Washington.

Wexler, Nancy. 1992. "Clairvoyance and Caution: Repercussions from the Human Genome Project." *The Code of Codes*, ed. D. Kevles and L. Hood, 211–43. Cambridge: Harvard University Press.

Whitney, C. 1997. "Europe Focuses on Future and the Impact of New Members." *New York Times*, 18 June, A3.

van Wijk, N. 1912. *Franck's Etymologisch Woordenboek der Nederlandsche Taal*. The Hague: Martinus Nijhoff.

Wilmut, Ian, et al. 1997. "Viable Offspring Derived from Fetal and Adult Mammalian Cells." *Nature* 385 (27 February): 810–13.

Yoder, Joella Gerstmeyer. 1988. *Unrolling Time: Christiaan Huygens and the Mathematization of Nature*. Cambridge: Cambridge University Press.

Young, Allan. 1982. "The Anthropologies of Illness and Sickness." *Annual Review of Anthropology* 11:257–85.

———. 1997. *Harmony of Illusions: Inventing Post-Traumatic Stress Disorder*. Princeton: Princeton University Press.

Zee, A. 1992. "Symmetry and the Search for Beauty in Modern Physics." *New Literary History* 23 (4): 815–38.

van der Zee, H. 1982. *The Hunger Winter: Occupied Holland, 1944–45*. London: Norman and Hobhouse.

INDEX

Karen-Sue Taussig is an associate professor of anthropology at the University of Minnesota.

Library of Congress Cataloging-in-Publication Data
Taussig, Karen-Sue, 1962–
Ordinary genomes : science, citizenship, and genetic identities /
Karen-Sue Taussig.
p. cm. — (Experimental futures : technological lives, scientific arts, anthropological voices)
Includes bibliographical references and index.
ISBN 978-0-8223-4516-9 (cloth : alk. paper)
ISBN 978-0-8223-4534-3 (pbk. : alk. paper)
1. Genetics—Social aspects—Netherlands. 2. Group identity—Netherlands.
3. National characteristics, Dutch. 4. Genetic counseling—Netherlands.
I. Title. II. Series: Experimental futures.
QH438.7.T38 2009
599.93′5—dc22
2009010550